I0057641

LIGHT AND VACUUM

The Wave-Particle Nature of the Light
and the Quantum Vacuum through the
Coupling of Electromagnetic Theory
and Quantum Electrodynamics

Constantin Meis

Institute for Nuclear Science & Technology, France

LIGHT AND VACUUM

The Wave-Particle Nature of the Light
and the Quantum Vacuum through the
Coupling of Electromagnetic Theory
and Quantum Electrodynamics

World Scientific

NEW JERSEY · LONDON · SINGAPORE · BEIJING · SHANGHAI · HONG KONG · TAIPEI · CHENNAI

Published by

World Scientific Publishing Co. Pte. Ltd.

5 Toh Tuck Link, Singapore 596224

USA office: 27 Warren Street, Suite 401-402, Hackensack, NJ 07601

UK office: 57 Shelton Street, Covent Garden, London WC2H 9HE

Library of Congress Cataloging-in-Publication Data
Meis, Constantin.
  Light and vacuum : the wave-particle nature of the light and the quantum vacuum through the coupling of electromagnetic theory and quantum electrodynamics / Constantin Meis (National Institute for Nuclear Science & Technology, France).
     pages cm
  Includes bibliographical references and index.
  ISBN 978-9814630894 (hardcover : alk. paper)
  1. Wave theory of light. 2. Vacuum. 3. Electromagnetic theory. 4. Quantum electrodynamics.
I. Title.
  QC670.M415 2014
  535'.15--dc23

                                           2014029163

British Library Cataloguing-in-Publication Data
A catalogue record for this book is available from the British Library.

Copyright © 2015 by World Scientific Publishing Co. Pte. Ltd.

All rights reserved. This book, or parts thereof, may not be reproduced in any form or by any means, electronic or mechanical, including photocopying, recording or any information storage and retrieval system now known or to be invented, without written permission from the publisher.

For photocopying of material in this volume, please pay a copying fee through the Copyright Clearance Center, Inc., 222 Rosewood Drive, Danvers, MA 01923, USA. In this case permission to photocopy is not required from the publisher.

In-house Editor: Song Yu

Typeset by Stalion Press
Email: enquiries@stallionpress.com

To my mother Aphrodite, to Angelika and Hypatia

"The infinite vacuum is the real essence of the cosmos, whatever exists has derived from the intrinsic action of the vacuum resulting to positive and negative entities"

(Anaximander, 611-546 BC).

Prologue

Throughout this book, we give the main principles of the electromagnetic theory and quantum electrodynamics (QED), both developed for the understanding of light's nature and for the explanation of the associated phenomena when interacting with matter. Of course, it is not in the scope of this manuscript to give a full and detailed presentation of these theories. Only selected theoretical topics have been chosen, supported by experimental evidence, which are indispensable for the understanding of the present status of the theories on the nature of light.

Furthermore, we discuss the main difficulties encountered by both theories, to ensure a complete and coherent mathematical description of the simultaneous wave-particle nature of light put in evidence by experiments.

Finally, we consider the basic aspect of QED related to the quantization of the vector potential amplitude of the electromagnetic field to a single photon state, and we advance elaborations on its relationship with the classical electromagnetic wave theory and the vacuum.

The topics are all drawn from many works previously published, and given in the bibliography. The perspectives and elaborations on the photon vector potential and its relationship to the quantum vacuum are of my own, in the aim to raise questions and aspire for further theoretical and experimental investigation, in order to improve our knowledge and understanding on the real essence of light and vacuum.

Note: In this book, the word "light" is not limited to the visible field, but concerns the whole electromagnetic spectrum ranging from zero frequency to infinity.

Table of Contents

Chapter 1

Introduction

"Light is the means nature employs to observe itself".

Man has always been fascinated by light. Since the rise of humanity, light is related to life and darkness to death. This concept seems to have intrinsic roots in the human mind whatever the tribe or nation, on any continent and during any historic or pre-historic period.

The natural entities emitting light, helping man and animals to see while eliminating the fears of darkness, like the sun, the moon and fire were venerated since the beginning and the absolute divine was associated with them. Man quickly understood that life, as it is met on the surface of this planet, would be inexistent or at least completely different without light. The forests are the main lags of the earth's atmosphere contributing to the production and to the recycling of oxygen, a so crucial for the life element. In the absence of the sun's light, plants, trees and forests are condemned to disappear and consequently animals would be unable to survive.

Within a biological point of view, light is the basic medium of nature conferring the possibility to organisms possessing an appropriate detection system, optical or not, to see or detect the shapes of the objects of the surrounding cosmos. By the same token, light is the means nature has chosen to watch itself by way of live beings. The exchange of energy and information between atoms, molecules and more complex systems occurs through light, permitting life to emerge in the cosmos. This notion, once understood, automatically attributes a unique character to light in the universe.

In the old historical times, light on earth was only emitted from fire, volcanic lava or high temperature metallic objects. Ancient

1

Greeks had such a profound respect for light that the only origin they could imagine was undoubtedly divine. In the Greek language, light is Phos ($\Phi\omega$s). This one-syllable word has quite a strong and deep consonance that used to be attributed only to exceptional physical entities, like Pyr ($\Pi\upsilon\rho$) fire, or even the head of the Gods, Zeus ($Z\varepsilon\upsilon$s). Hence, according to the legend, Prometheus has stolen the fire from Gods in order to offer light and heat to mankind. Disregarding the fact if mankind was worthy of it, the value of the gift was invaluable. For his action, Prometheus was punished by the Gods to be attached on the Caucasian mountains where an eagle was eating his liver in the day while it was regenerated during the night. The character of this endless torture without any chance of forgiveness or clemency gives witness to the extreme importance ancient Greeks attributed to the stolen entity from the Gods, fire. One simply has to wonder what human civilization would be like without fire.

Not surprisingly, *in most religions God is identified with light.*

However, light has an intrinsic nature and undergoes specific physical laws governing its behavior, and studied by science. Today, we perfectly understand the behavior laws but our mathematical representation of its wave-particle nature and its relationship to the vacuum is still incomplete.

Chapter 2

Historical Survey
and Experimental Evidence

The concepts of light during the last 2500 years: corpuscles, ray optics, wave optics, electromagnetic wave theory and finally quantum particle theory.

As far as we know, the first attempts to study the nature of light from the scientific point of view are due to ancient Greeks. They believed light to be composed of corpuscles.

Thales of Miletus in the sixth century BC knew already that in a given medium, light propagates in straight lines and that the light of the sun also obeys that property. Based on that knowledge and using his famous theorem, he measured the height of the Cheops pyramid by comparing the length of its shadow to that of his stick positioned vertically.

About two centuries later, Euclid published the book "Optica" in which, based on the rectilinear propagation of light, he developed the laws of reflection by applying principles of geometry. Archimedes, developed the geometrical study of parabolas and ellipsoids and, according to the legend, he created big metallic mirrors for focusing the sunlight in order to burn the Roman battle ships during the siege of Syracuse. That was the first time in human history that light was employed as a weapon.

A couple of centuries later, in his book "Optics", Ptolemy of Alexandria made a full synthesis of all the previous knowledge on light following the concepts of Euclid, Archimedes and Heron of Alexandria, treating refraction, reflection and colors. Refraction of the moonlight and sunlight by the earth's atmosphere was also analyzed.

Euclid's "Optica" and Ptolemy's "Optics" are the first known scientific publications on light's properties. Only a fraction of "Optics" has been saved, and this is thanks to Arab mathematicians of the 7th to 10th centuries AC.

After the dominance of Christians during the third century AC, the first Byzantine emperors ordered the definite closure and destruction of all Greek mathematical and philosophical academies that were spread from Greece to Egypt through the Middle East. Philosophers and scientists were persecuted and murdered preventing any possibility of scientific development for more than thirteen centuries. The last one of them was Hypatia of Alexandria.

Humanity had to wait until 1620 AC for Snell's works on refraction and Fermat's principle, according to which light rays travel along the path of least time. It was at this period that many scientists like, Grimaldi, Boyle, Hook, Descartes and others began studying light's properties.

Half a century later, in the year of 1670, Newton had retaken the proposals of Pierre Gassendi, who had revived the ideas of ancient Greeks, and advanced the theory that light rays are composed of corpuscles that travel rectilinearly. However, he went further beyond this description by announcing that under specific physical conditions, the light corpuscles may give birth to waves in "aether", without giving it a precise definition. Surprisingly, in the first edition of his book "Opticks", the light corpuscles and the generated "aether" waves were replaced by particles submitted to a kind of "periodic relief".

Parallel to Newton, in the year of 1680, Huygens developed a remarkable wave theory for light, deducing the laws of reflection and refraction while demonstrating that wave propagation may not be in contradiction with the rectilinear propagation. Huygens' wave theory was a hard opponent to Newton's particles concept.

In the beginning of the 19th century, Young's experiments demonstrated that interference can be obtained by different waves, while Fresnel applied the wave theory to explain the diffraction patterns observed experimentally. Furthermore, Young explained some polarization observations by making the hypothesis that light oscillations are perpendicular to the propagation axis. Nevertheless, not even a scientist could seriously consider the interference theory and Young's studies were berated by the journalists. Nearly twenty years later, the diffraction theories of Young and Fresnel, though not identical, were the only ones capable of predicting all the observed diffraction patterns and that was really the crucial turnover point in the 19th century following which the scientific community started to generally accept the wave nature of light.

The wave theory was dominant until the beginning of 20th century. It is worth noting that until that time, nearly for more than 2500 years, the main question for scientists was focused on the particle or the wave nature of light, but nobody had advanced any specific questions on the real nature of light, i.e., a corpuscle made of what? or a wave of what?

In 1865, James Clerk Maxwell published his remarkable work on the electromagnetic waves issued from ... Maxwell's equations describing the relations between the electric and magnetic fields, and has shown that the electromagnetic waves propagate in vacuum at the same speed observed by astronomers for light. For the first time the speed of light was related directly to the vacuum electric permittivity and magnetic permeability showing the natural relationship between light and vacuum. Just a few years later, Hertz discovered the long wavelength electromagnetic radiation demonstrating that it propagates at the speed of light confirming Maxwell's theory.

It is quite amazing in the history of science that Maxwell's and Hertz's works, carried out within roughly ten years, were absolutely revolutionary and decisive in our understanding about the nature of light, which remained stationary for over 25 centuries.

At the same period, from 1880 to 1900, the works of Stefan, Wien and Rayleigh have shown for the first time the direct relationship between the thermal radiation energy and the temperature of the emitting body, assimilated to a black body. However, the emitted radiation energy density of the black body as a function of the temperature calculated by Rayleigh, failed to describe the experimental results obtained at short wavelengths. Scientists had given the name of "*UV catastrophe*" to this situation which revealed the necessity of a new theoretical approach.

In the very beginning of the 20th century, Max Planck assumed that the bodies are composed of "resonators", a kind of oscillators that have the particularity of emitting the electromagnetic energy in packets of $h\nu$, thus proportional to the wave frequency ν times h, that was later called Planck's constant. This hypothesis permitted Planck to establish the correct energy density expression for the radiation emitting from a black body with respect to temperature, which is in excellent agreement with the experiment.

In 1902, Lenard pointed out that the photoelectric effect, discovered by Hertz fifteen years earlier, occurs beyond a threshold frequency of light and the kinetic energy of the emitted electrons does not depend on the incident light intensity.

The experiments carried out by Michelson and Morley demonstrated that the speed of light in vacuum is a universal physical constant. This fundamental discovery was the starting point for the development of the theory of special relativity, based on Lorentz's set of equations followed by Poincare's theoretical studies. Einstein grouped all these works in an elaborated theory published in 1905. During the same year, based on the works of Planck and Lenard, Einstein published also an article stating that the electromagnetic radiation is composed of quanta of energy $h\nu$ and advanced a simple particle interpretation of the photoelectric effect. In a second article in 1917, Einstein re-established Planck's radiation density formula and expressed that the light quanta have a momentum $h\nu/c$, where c is the velocity of light in vacuum. He advanced that "*the energy of a light ray when spreading from a point consists of a finite number of energy quanta localized in points in space, which move without dividing and are only absorbed and emitted as a whole*". That was a decisive step toward the particle theory of light, but Bohr, who was strongly opposed to the photon concept, announced in his Nobel lecture, "*the light quanta hypothesis is not compatible with the interference phenomena and consequently it cannot throw light in the nature of radiation*". The concept of light composed of quanta was still not generally accepted.

Compton published his works on X-ray scattering by electrons in 1923 showing that the experimental results can only be interpreted based on the light quanta model. On the other side, Wentzel in 1926 and Beck in 1927 demonstrated that *"the photoelectric effect may be quite well interpreted using Maxwell's wave theory for radiation and quantum theory for the atomic energy levels without ever having to introduce the light quanta"* in their calculations. Although their articles were an extremely strong argument in favor of the wave theory, it is quite surprising that exactly the same year, the light quanta concept started to be universally accepted by many scientists and Lewis introduced the word "Photon", from the Greek word Phos ($\Phi\omega s$ = light).

Obviously, just before 1930, the general picture on the nature of light was extremely puzzling and confusing since both opposing sides defending the wave or the particle theory presented equally strong arguments. Not surprisingly, in 1928, Bohr inspired by De Broglie's thesis on the simultaneous wave character of particles announced the *"Complementarity Principle"* according to which *"light has both wave and particle natures appearing mutually exclusive in each specific experimental condition"*.

From 1925 to 1930, new theories have shaken physics. Heisenberg developed *matrix-mechanics*, while in parallel, Schrödinger invented *wave-mechanics* which was revealed to be equivalent to *matrix-mechanics*. Dirac and Jordan developed the *quantum theory of radiation*. A few years later, enhancing further Dirac's works, Pauli and Heisenberg developed the fundamentals of *quantum electrodynamics* (QED) theory which was further developed in the years of forties to sixties. According to this theory, photons are considered to be *point particles* and the energy of the electromagnetic field corresponds to, that of an ensemble of quantized harmonic oscillators. In the complete absence of photons, QED representation gives an infinite energy for the vacuum state.

By the end of 20th century and the beginning of the 21st, astrophysical observations established the first estimations of the vacuum energy in the universe demonstrating that it is measurable. In 1989, Weinberg argued that when considering reasonable cut-offs in the

UV spectrum for the energy of high frequency photons, the vacuum energy density obtained by QED is by far many orders of magnitude greater than the measured one. This has been recently called the *"quantum vacuum catastrophe"* in order to mark the analogy with the *"UV catastrophe"* which, a century ago, entailed the necessity of developing a new theory to account with.

Considering that the fundamentals of QED were established in the years of 1930 to 1950, since then no new theories came to light. Nevertheless, the technological advances on microwaves during the Second World War and the invention of masers and lasers in the years of the fifties and sixties, respectively, gave scientists an impulse for advanced experiments on light's nature. Some of them are briefly described in this book.

Bibliography

1. A.I. Akhiezer and B.V. Berestetskii, *Quantum electrodynamics*, New York: Interscience Publishers. 1965.
2. C.L. Andrews, *Optics of the electromagnetic spectrum*, Englewood cliffs, New Jersey: Prentice-Hall Inc. 1960.
3. G. Auletta, *Foundations and interpretation of quantum mechanics*, Singapore: World Scientific. 2001.
4. G. Beck, Zur theorie des photoeffekts, *Z. Phys.* **41** (1927) 443–452.
5. N. Bohr, H.A. Kramers and J.C. Slater, The quantum theory of radiation, *Phil. Mag.* **47**(281) (1924) 785–802.
6. M. Born and E. Wolf, *Principles of optics,* Cambridge: Harvard University Press. 1999.
7. S. Bourzeix, B. de Beauvoir, F. Bez, M.D. Plimmer, F. de Tomasi, L. Julien, F. Biraben and D.N. Stacey, High resolution spectroscopy of the hydrogen atom: determination of the 1S Lamb shift, *Phys. Rev. Lett.* **76** (1996) 384.
8. W. Bothe and H. Geiger, Über das wesen des compton effekts: ein experiment ellerbeitrag zur theories der strahlung, *Z. Phys.* **32**(9) (1925) 639–663.
9. B.H. Bransden and C.J. Joachain, *Physics of atoms and molecules*, London: Longman Group Ltd. 1983.
10. A.H. Compton, The spectrum of scattered x-rays, *Phys. Rev.* **22**(5) (1923) 409–413.
11. L. de Broglie, *The revolution in physics; a non-mathematical survey of quanta*, New York: Noonday Press. 1953, pp. 117, 178–186.
12. L. de Broglie, *Une tentative d'interprétation causale et non linéaire de la mécanique ondulatoire: la théorie de la double solution*, Paris: Gauthier Villars. 1956.
13. P.A.M. Dirac, *The principles of quantum mechanics*, Oxford: Oxford University Press. 1958.

14. A. Einstein, *The collected papers of Albert Einstein*, J. Stachel, D.C. Cassidy, J. Renn, & R. Schulmann (Eds.) Princeton, New Jersey: Prinston University Press. 1987.
15. R. Feynman, *The strange theory of light and matter*, Princeton, New Jersey: Princeton University Press. 1988.
16. J.C. Garrison and R.Y. Chiao, *Quantum optics*, Oxford: Oxford University Press. 2008.
17. H. Haken, *Light*, Amsterdam–Oxford: North Holland Publishing. 1981.
18. W. Heitler, *The quantum theory of radiation*, Oxford: Clarendon Press. 1954.
19. T. Hey, *The new quantum universe*, Cambridge: Cambridge University Press. 2003.
20. S. Jeffers, S. Roy, J.P. Vigier and G. Hunter, *The present status of the quantum theory of light*, Boston: Kluwer Academic Publishers. 1997.
21. W.E. Lamb Jr and M.O. Scully, In *Polarization, matière et rayonnement, volume jubilaire en l' honneur d' Alfred Kastler*, French Physical Society, Paris: Press Universitaires de France. 1969.
22. H.A. Lorentz, Simplified theory of electrical and optical phenomena in moving systems, *Proceedings, The Royal Netherlands academy of art and sciences* **1** (1899) pp. 427–442.
23. H.A. Lorentz, Electromagnetic phenomena in a system moving with any velocity smaller than that of light, *Proceedings, The Royal Netherlands academy of art and sciences* **6** (1904) pp. 809–831.
24. A.A. Michelson, H.A. Lorentz, D.C. Miller, R.J. Kennedy, E.R. Hedrick and P.S. Epstein, The Smithsonian/NASA astrophysics data system; conference on the Michelson-Morley experiment held at Mount Wilson, February, 1927, *Astrophys. J.* **68** (1928) 341–402.
25. A.R. Mickelson, *Physical optics*, Cleveland, Ohio: Van Nostrand Reinhold. 1992.
26. P.W. Milonni, *The quantum vacuum*, London: Academic Press Inc. 1994.
27. M.H. Mittleman, *Introduction to the theory of laser-atom interactions*, New York: Plenum Press. 1982.
28. I. Newton, *Opticks*, New York: Dover Publications. 1952.
29. M. Planck, *The theory of heat radiation*, New York: Dover Publications. 1959.
30. F.H. Read, *Electromagnetic radiation*, New York: John Wiley & Sons. 1980.
31. V. Ronchi, *The nature of light: an historical survey*, Cambridge: Harvard University Press. 1970.
32. A. Rougé, Relativité restreinte, la contribution d'Henri Poincaré, *Editions Ecole Polytechnique* (2008) 135.
33. L.H. Ryder, *Quantum field theory*, Cambridge: Cambridge University Press. 1987.
34. B.E.A. Saleh and M.C. Teich, *Fundamentals of photonics*, New York: John Wiley & Sons. 2007.
35. J. Schwinger, L.L. DeRaad Jr. and K.A. Milton, Casimir effect in dielectrics, *Ann. Phys. (N.Y.)* **115**(1) (1978).
36. G.I. Taylor, Interference fringes with feeble light, *Proc. Cam. Philos. Soc.* **15** (1909) 114–115.

37. P.H.G.M. van Blokland and J.T.G. Overbeek, *J. Chem. Soc. Faraday Trans.* **74** (1978) 2637–2651 (1978).
38. L.J. Wang, X.Y. Zou and L. Mandel, Experimental test of the de Broglie guided-wave theory for photons, *Phys. Rev. Lett.* **66** (1991) 1111.
39. S. Weinberg, The cosmological constant problem, *Rev. Mod. Phys.* **61** (1989) 1.
40. M. Weissbluth, *Photon-atom interactions*, London: Academic Press Inc. 1988.
41. G. Wentzel, Zur theorie des photoelektrischen effekts, *Z. Phys.* **40** (1926) 574–589.
42. A.K. Zvzdin and V.A. Kotov, *Modern magneto-optics and magneto-optical materials*, Bristol and Philadelphia: Institute of Physics Publishing. 1997.

Chapter 3

Basic Principles
of the Electromagnetic
Wave Theory

3.1. Maxwell's Equations

The works of Ampere and Faraday on the electric and magnetic fields during the years of the 1860s were the starting point for establishing the theory of electromagnetic waves.

Initially they were both investigating time varying electric and magnetic fields. Faraday firstly understood that the curl of an electric field $\vec{E}\,(\vec{r}, t)$ (units: Volt m^{-1}) at a position $\vec{r}$ and at an instant t, in a medium with magnetic permeability μ (units: Henry m^{-1}) equals the time variation of the magnetic field intensity $\vec{H}\,(\vec{r}, t)$ (units: Ampere m^{-1}) times μ

$$\vec{\nabla} \times \vec{E}\,(\vec{r}, t) = -\mu \frac{\partial}{\partial t} \vec{H}\,(\vec{r}, t) \qquad (3.1.1)$$

The last equation is still called Faraday's law. A few years later, Ampere put in evidence that in a medium with electric permittivity ε (units: Farad m^{-1}), the curl of the magnetic field intensity $\vec{H}\,(\vec{r}, t)$ equals the current density $\vec{j}\,(\vec{r}, t)$ plus the time variation of the electric field intensity $\vec{E}\,(\vec{r}, t)$ times ε.

$$\vec{\nabla} \times \vec{H}\,(\vec{r}, t) = \vec{j}\,(\vec{r}, t) + \varepsilon \frac{\partial}{\partial t} \vec{E}\,(\vec{r}, t) \qquad (3.1.2)$$

where the current density $\vec{j}\,(\vec{r}, t)$ (Coulomb m^{-2} s^{-1} or Ampere m^{-2}) is related to the charge density $\rho\,(\vec{r}, t)$ (Coulomb m^{-3}) through the fundamental "continuity equation"

$$\vec{\nabla} \cdot \vec{j}(\vec{r}, t) = -\frac{\partial \rho(\vec{r}, t)}{\partial t} \qquad (3.1.3)$$

expressing that the spatial variation of the current density equals the time variation of the charge density implying simply charge conservation.

What one first notices is the natural fact that the electric and magnetic fields cannot evaluate independently of the medium which is expressed in this case through the electric permittivity and the magnetic permeability. We can thus introduce the notions of the electric displacement flux density $\vec{D}(\vec{r}, t)$ (units: Coulomb m^{-2}) and magnetic field flux density, that is the magnetic induction, $\vec{B}(\vec{r}, t)$ (units: Webers m^{-2}) which are related to the electric and magnetic fields intensities $\vec{E}(\vec{r}, t)$ and $\vec{H}(\vec{r}, t)$ respectively through the tensor expressions,

$$\vec{D}_{\alpha\beta}(t) = \hat{\varepsilon}_{\alpha\beta} \cdot \vec{E}_{\alpha\beta}(t); \quad \vec{B}_{\alpha\beta}(t) = \hat{\mu}_{\alpha\beta} \cdot \vec{H}_{\alpha\beta}(t) \qquad (3.1.4)$$

where the electric permittivity and magnetic permeability respectively are tensors $\hat{\varepsilon}_{\alpha\beta}$ and $\hat{\mu}_{\alpha\beta}$ with $\alpha = x, y, z$ and $\beta = x, y, z$, characterizing the intrinsic electric and magnetic nature of the medium in which the electric and magnetic field intensities subsist.

At that point it is important mentioning that the electric field $\vec{E}(\vec{r}, t)$ and the magnetic induction $\vec{B}(\vec{r}, t)$ are fundamental fields while $\vec{D}(\vec{r}, t)$ and $\vec{H}(\vec{r}, t)$ are fields that include the response of the medium at macroscopic level.

In isotropic media, i.e., in media in which the electric and magnetic properties are identical in all spatial directions, the electric field is parallel to the electric displacement and the magnetic field flux intensity parallel to the magnetic field

$$\vec{D}(\vec{r}, t) = \varepsilon \vec{E}(\vec{r}, t) \quad \vec{H}(\vec{r}, t) = \frac{1}{\mu} \vec{B}(\vec{r}, t) \qquad (3.1.5)$$

Taking the divergence ($\vec{\nabla} \cdot$) of the Faraday's and Ampere's laws, (3.1.1) and (3.1.2), and using the continuity equation (3.1.3) one gets directly Gauss's laws

$$\vec{\nabla} \cdot \vec{E}\,(\vec{r}, t) = \frac{\rho\,(\vec{r}, t)}{\varepsilon} \tag{3.1.6}$$

$$\vec{\nabla} \cdot \vec{H}\,(\vec{r}, t) = 0 \tag{3.1.7}$$

where (3.1.6) is also called Coulomb's law.

Physically speaking Gauss's equations express the fact that individual electric charges, positive as well as negative, subsist in nature and are expressed in the charge density $\rho\,(\vec{r}, t)$, while the absence of the magnetic monopoles is expressed in equation (3.1.7).

James Clerk Maxwell in 1873 unified equations (3.1.1), (3.1.2), (3.1.6) and (3.1.7) to a group called Maxwell's equations which, joined to the continuity equation (3.1.3) express the relationship between the electric and magnetic fields intensities in a medium with a given magnetic permeability μ and electric permittivity ε in the presence of charges and current densities.

Frequently, Maxwell's equations are written in such a way that they express the relation between the electric field intensity $\vec{E}\,(\vec{r}, t)$ and the magnetic field flux intensity $\vec{B}\,(\vec{r}, t)$, i.e., the fundamental fields.

$$\vec{\nabla} \times \vec{E}\,(\vec{r}, t) = -\frac{\partial}{\partial t}\vec{B}\,(\vec{r}, t) \tag{3.1.8}$$

$$\vec{\nabla} \times \vec{B}\,(\vec{r}, t) = \mu \vec{j}\,(\vec{r}, t) + (\varepsilon\mu)\frac{\partial}{\partial t}\vec{E}\,(\vec{r}, t) \tag{3.1.9}$$

$$\vec{\nabla} \cdot \vec{E}\,(\vec{r}, t) = \frac{\rho\,(\vec{r}, t)}{\varepsilon} \tag{3.1.10}$$

$$\vec{\nabla} \cdot \vec{B}\,(\vec{r}, t) = 0 \tag{3.1.11}$$

The relations (3.1.8) to (3.1.11) together with the continuity equation (3.1.3) are the fundamental equations governing the behavior of the electromagnetic fields. They are invariant under time reversal $(t \to -t)$, or space inversion $(\vec{r} \to -\vec{r})$ or even both time reversal and space inversion. More precisely, the intrinsic space time transformation properties are such that the electric field $\vec{E}\,(\vec{r}, t)$, and therefore the electric displacement $\vec{D}\,(\vec{r}, t)$, do not change sign under time reversal but under space inversion. The opposite is valid for the magnetic induction $\vec{B}(\vec{r}, t)$ and the magnetic field $\vec{H}(\vec{r}, t)$. This is also true for the response of a medium to an electric field and a magnetic induction consisting respectively by the electric polarization $\vec{P}(\vec{r}, t)$ (units: Coulomb m^{-2}) and magnetization $\vec{M}(\vec{r}, t)$ (units: Ampere m^{-1}) expressed as following

$$\vec{P}(\vec{r}, t) = \vec{D}(\vec{r}, t) - \varepsilon_0 \vec{E}(\vec{r}, t) \tag{3.1.12}$$

$$\vec{M}(\vec{r}, t) = \frac{1}{\mu_0} \vec{B}(\vec{r}, t) - \vec{H}(\vec{r}, t) \tag{3.1.13}$$

According to the set of equations (3.1.8) to (3.1.11), in space regions characterized by the absence of charges and current densities $|\vec{j}| = 0$, $\rho = 0$, the physical vector entities $\vec{E}(\vec{r}, t)$ and $\vec{B}(\vec{r}, t)$ are related through the quantities ε and μ which appear to be of crucial importance in Ampere's equation (3.1.9). Under these conditions Maxwell's equations get the simple form

$$\vec{\nabla} \times \vec{E}(\vec{r}, t) = -\frac{\partial}{\partial t} \vec{B}(\vec{r}, t) \tag{3.1.14}$$

$$\vec{\nabla} \times \vec{B}(\vec{r}, t) = (\varepsilon\mu)\frac{\partial}{\partial t} \vec{E}(\vec{r}, t) \tag{3.1.15}$$

$$\vec{\nabla} \cdot \vec{E}(\vec{r}, t) = 0 \tag{3.1.16}$$

$$\vec{\nabla} \cdot \vec{B}(\vec{r}, t) = 0 \tag{3.1.17}$$

Starting from the last set of equations, Maxwell deduced the time varying propagation of the electric and magnetic fields that gave birth, for the first time, to a theoretical representation of the electromagnetic waves.

3.2. Electromagnetic Wave Propagation

• Dispersion relation

Let us analyze how Maxwell's equations lead naturally to the expressions of propagation of the electromagnetic fields. We make the hypothesis that the electric field intensity $\vec{E}$ and the magnetic field flux intensity $\vec{B}$ at any spatial position $\vec{r}$ vary in time with an angular frequency $\omega = 2\pi\nu = 2\pi\frac{1}{T}$, where ν and T are the frequency and the period of the oscillation respectively, so as we can write

$$\begin{bmatrix} \vec{E}\,(\vec{r},t) \\ \vec{B}\,(\vec{r},t) \end{bmatrix} = Re \left\{ \begin{bmatrix} \vec{E}_0\,(\vec{r}) \\ \vec{B}_0\,(\vec{r}) \end{bmatrix} e^{i\omega t} \right\} \tag{3.2.1}$$

where Re denotes the real part of the equation.

Then, Maxwell's equations, more precisely Ampere's and Faraday's laws, are expressed through the angular frequency ω

$$\vec{\nabla} \times \vec{E}\,(\vec{r},t) = i\omega\,\vec{B}\,(\vec{r},t) \tag{3.2.2}$$

$$\vec{\nabla} \times \vec{B}\,(\vec{r},t) = -i\omega\,(\varepsilon\mu)\vec{E}\,(\vec{r},t) \tag{3.2.3}$$

$$\vec{\nabla} \cdot \vec{E}\,(\vec{r},t) = 0 \tag{3.2.4}$$

$$\vec{\nabla} \cdot \vec{B}\,(\vec{r},t) = 0 \tag{3.2.5}$$

The set of the above equation describes the behavior of the electric field and magnetic field flux intensities of an electromagnetic harmonic wave oscillating with an angular frequency ω in an isotropic medium with electric permittivity ε and magnetic permeability μ.

One should notice two important points:

a. the correspondence $\frac{\partial}{\partial t} \rightarrow -i\omega$, which as we shall see, will be of crucial importance in quantum mechanics formalism,
b. the curl inversion absence of symmetry between $\vec{E}$ and $\vec{B}$ which is ensured through the factor $-(\varepsilon\mu)$, and which is revealed to be of high importance in the electromagnetic field theory.

The first remark will be extensively discussed in the next chapter. However, the second reveals the role of the medium's nature in the propagation of the electromagnetic field.

Let us assume the media to be homogeneous. Therefore we can use plane wave solutions for $\vec{E}_0\,(\vec{r})$ and $\vec{B}_0\,(\vec{r})$ in (3.2.1) of the form $e^{i\vec{k}\cdot\vec{r}}$ where $\vec{k}$ is a vector along the propagation axis called wave vector. Hence, from Maxwell's equations we obtain

$$\vec{k} \times \vec{E}\,(\vec{r},t) = \omega\,\vec{B}\,(\vec{r},t) \tag{3.2.6}$$

$$\vec{k} \times \vec{B}\,(\vec{r}, t) = -\omega\,(\mu\varepsilon)\vec{E}\,(\vec{r}, t) \tag{3.2.7}$$

$$\vec{k} \cdot \vec{E}\,(\vec{r}, t) = 0 \tag{3.2.8}$$

$$\vec{k} \cdot \vec{B}\,(\vec{r}, t) = 0 \tag{3.2.9}$$

The last set of equations entail that in isotropic media $\vec{k}$ is perpendicular to $\vec{E}$ and $\vec{B}$ (while it is always perpendicular to $\vec{D}$ and $\vec{B}$, even if the medium is anisotropic) and $\vec{E}$ is perpendicular to $\vec{B}$.

Consequently, the unit vectors, $\hat{e} = \frac{\vec{E}}{|\vec{E}|}$, $\hat{b} = \frac{\vec{B}}{|\vec{B}|}$, $\hat{k} = \frac{\vec{k}}{|\vec{k}|}$ constitute a right-hand rectangular coordinate system.

Considering equation (3.2.6) and taking the cross-product we get

$$\vec{k} \times \vec{k} \times \vec{E}\,(\vec{r}, t) = \omega\,\vec{k} \times \vec{B}\,(\vec{r}, t) \tag{3.2.10}$$

According to the vector property $\vec{k} \times \vec{k} \times \vec{E}\,(\vec{r}, t) = (\vec{k} \cdot \vec{E}\,(\vec{r}, t))\vec{k} - k^2 \vec{E}\,(\vec{r}, t) = -k^2 \vec{E}\,(\vec{r}, t)$ where we have also used equation (3.2.8), the equation (3.2.10) using (3.2.7) becomes

$$\Omega_k^2 \vec{E}\,(\vec{r}, t) = 0 \tag{3.2.11}$$

with $\Omega_k^2 = k^2 - (\mu\varepsilon)\,\omega^2$. The solution to the last equation is $\Omega_k^2 = 0$. Thus,

$$k^2 = (\mu\varepsilon)\,\omega^2 \tag{3.2.12}$$

which is called the "dispersion relation" relating the wave vector k to the angular frequency ω through the medium intrinsic properties ε and μ.

- Physical quantities involved in the electromagnetic field

Notice that $\frac{1}{\varepsilon\mu}$ has the dimension of the square power of celerity. Consequently, it is precisely the electric permittivity ε and magnetic permeability μ that fixes the velocity of the electromagnetic wave in a medium whose refractive index n is defined as

$$n = \sqrt{\frac{\mu\varepsilon}{\mu_0\varepsilon_0}} \tag{3.2.13}$$

with $\mu_0 = 4\pi \times 10^{-7}\mathrm{H\,m^{-1}}$ and $\varepsilon_0 = 8.85 \times 10^{-12}\mathrm{F\,m^{-1}} \approx \frac{1}{36\pi} 10^{-9}\mathrm{F\,m^{-1}}$, the magnetic permeability and electric permittivity of the vacuum.

This is already an amazing concept of the classical electromagnetic theory that confers electric and magnetic properties to the vacuum state.

The characteristic impendence of a medium is given by

$$R = \sqrt{\frac{\mu}{\varepsilon}} \tag{3.2.14}$$

corresponding to the resistance (units: Ohms, Ω) of the medium when an electric potential is applied.

Consequently, the vacuum itself is characterized by the impendence

$$R_{\text{vacuum}} = \sqrt{\frac{\mu_0}{\varepsilon_0}} \approx 120\pi \ \Omega \tag{3.2.15}$$

which has the definite value of 376.7 Ω.

It is extremely worthy noticing that the vacuum state in classical electrodynamics is not an empty, fully inertial entity characterized by the perfect "inexistence" but has its own physical, measurable impendence, R_{vacuum}, as well as electric, ε_0, and magnetic, μ_0, properties affecting the propagation of the electromagnetic waves in vacuum. In fact, the speed of the electromagnetic waves in vacuum, c, results directly from the intrinsic properties of the vacuum itself

$$c = \frac{1}{\sqrt{\varepsilon_0 \mu_0}} \approx 2.9979 \ 10^8 \ \text{m s}^{-1} \tag{3.2.16}$$

Although the origin of the vacuum's electromagnetic properties remains unknown, the last ones dictate the value of the highest possible velocity in the universe, c, in the frame of course of the present state of knowledge.

The dispersion relation writes now

$$k^2 \left(\frac{c^2}{n^2} \right) = \omega^2 \tag{3.2.17}$$

expressing a fundamental relationship between the wave vector k and the angular frequency ω through the speed of the light c/n in a medium with refractive index n, where $n = 1$ for the vacuum.

- Propagation equations of electromagnetic waves

   Taking the curl of equation (3.1.14) and using (3.1.15)

$$\vec{\nabla} \times \vec{\nabla} \times \vec{E}\,(\vec{r}, t) = -\frac{\partial}{\partial t}\,(\vec{\nabla} \times \vec{B}(\vec{r}, t)) = -\frac{\partial}{\partial t}\,(\mu\varepsilon)\frac{\partial}{\partial t}\vec{E}\,(\vec{r}, t)$$

$$= -(\mu\varepsilon)\frac{\partial^2}{\partial t^2}\vec{E}\,(\vec{r}, t) \qquad (3.2.18)$$

   Considering the classical vector formalism

$$\vec{\nabla} \times \vec{\nabla} \times \vec{E} = \vec{\nabla}(\vec{\nabla} \cdot \vec{E}) - \vec{\nabla}^2\vec{E} = -\vec{\nabla}^2\vec{E} \qquad (3.2.19)$$

So (3.2.18) writes

$$\vec{\nabla}^2\vec{E}\,(\vec{r}, t) - (\mu\varepsilon)\frac{\partial^2}{\partial t^2}\vec{E}\,(\vec{r}, t) = 0 \qquad (3.2.20)$$

which is the propagation equation for the electric field in a medium with refractive index n

$$\vec{\nabla}^2\vec{E}\,(\vec{r}, t) - \frac{n^2}{c^2}\frac{\partial^2}{\partial t^2}\vec{E}\,(\vec{r}, t) = 0 \qquad (3.2.21)$$

   An analogue equation can be obtained for the magnetic field flux intensity $\vec{B}\,(\vec{r}, t)$.

- Helmholtz equation

   The vector relations of (3.2.6) to (3.2.9) show that $\vec{k}$, $\vec{E}$ and $\vec{B}$ are perpendicular to each other and construct a right-hand coordinate system. Hence, the harmonic solutions of the propagation equation for the electric field and magnetic field induction write

$$\vec{E}\,(\vec{r}, t) = E_0\,(\vec{r})\,\hat{e}\,e^{-i(\vec{k}\cdot\vec{r}-\omega t)} + c.c. \qquad (3.2.22)$$

$$\vec{B}\,(\vec{r}, t) = B_0\,(\vec{r})(\hat{k} \times \hat{e})\,e^{-i(\vec{k}\cdot\vec{r}-\omega t)} + c.c. \qquad (3.2.23)$$

with *c.c.* denoting the complex conjugate part.

   From (3.2.12), (3.2.17), (3.2.21) and (3.2.22), we deduce the Helmholtz equation for each component of the electric field

$$\left(\frac{\partial^2}{\partial x^2} + \frac{\partial^2}{\partial y^2} + \frac{\partial^2}{\partial z^2} + k^2\right) E_\alpha = 0 \qquad \alpha = (x, y, z) \qquad (3.2.24)$$

In other words, in the classical electromagnetic theory, the electric and magnetic fields oscillate perpendicularly along the propagation axis k with an angular frequency ω at the speed c/n.

This is called simply a plane wave solution. In this representation the wave vector in vacuum ($n = 1$) has the simple expression

$$\vec{k} = \frac{\omega}{c}\hat{k} = \frac{2\pi\,\nu}{c}\hat{k} = \frac{2\pi}{\lambda}\hat{k} \tag{3.2.25}$$

where λ is the wavelength, representing the spatial distance of a complete oscillation of the electric and magnetic field vectors over a period T and $\hat{k}$, the unit vector along the propagation axis.

For a given fixed value of the angular frequency ω, and the wavelength λ, the wave is called monochromatic (single color) and we will use the notations $\vec{E}_\omega\,(\vec{r}, t)$ and $\vec{H}_\omega\,(\vec{r}, t)$ for the corresponding electric and magnetic fields, which, according to the above considerations obey the same propagation equation

$$\vec{\nabla}^2\vec{G}_\omega(\vec{r}, t) - \frac{n^2}{c^2}\frac{\partial^2}{\partial t^2}\vec{G}_\omega(\vec{r}, t) = 0 \tag{3.2.26}$$

where $\vec{G}_\omega(\vec{r}, t)$ is either $\vec{E}_\omega\,(\vec{r}, t)$ or $\vec{H}_\omega\,(\vec{r}, t)$.

More generally an electromagnetic wave is called Transverse Electro-Magnetic (TEM) when both $\vec{E}_\omega(\vec{r}, t)$ and $\vec{H}_\omega(\vec{r}, t)$ lie in the plane that is transverse to the direction of propagation, with the same token Transverse Electric (TE) corresponding to the configuration with all the components of the electric field, but not those of the magnetic field, in the transverse plane to the propagation axis. Conversely, a Transverse Magnetic (TM) electromagnetic wave has all of the components of the magnetic field, but not those of the electric one, in the plane transverse to the propagation direction.

• Energy flux of electromagnetic waves

The vector product

$$\vec{S}_\omega = \vec{E}_\omega \times \vec{H}_\omega \tag{3.2.27}$$

represents the instantaneous magnitude and direction of the power flow of the electromagnetic field and is called the Poynting vector.

The power density (units: Watt m^{-2}) along the propagation axis is obtained by time averaging the Poynting vector over a period T

$$\langle \vec{S}_\omega \rangle_T = \frac{1}{T} \int_0^T (\vec{E}_\omega \times \vec{H}_\omega)\, dt = \frac{1}{2}\mathrm{Re}\left\{ \frac{\vec{E}_\omega \times \vec{B}_\omega}{\mu} \right\}$$

$$= \frac{1}{2}\left(\frac{c}{n}\right)\varepsilon|\vec{E}_\omega|^2\,\hat{k} = \frac{1}{2}\left(\frac{c}{n}\right)\frac{1}{\mu}|\vec{B}_\omega|^2\hat{k} \qquad (3.2.28)$$

The quantity $\langle \vec{S}_\omega \rangle_T$ is also called "Intensity" or "Irradiance" of the monochromatic electromagnetic wave.

The energy density (units: Joules m^{-3}) of a transverse monochromatic electromagnetic (TMEM) wave in vacuum oscillating at the angular frequency ω is obtained at the coordinate $\vec{r}$ and at an instant t by the expression:

$$W_\omega(\vec{r},t) = \frac{1}{2}\left(\varepsilon_0|\vec{E}_\omega(\vec{r},t)|^2 + \frac{1}{\mu_0}|\vec{B}_\omega(\vec{r},t)|^2 \right) \qquad (3.2.29)$$

where ε_0 and μ_0 are the vacuum permittivity and permeability related to the velocity of light in vacuum c by the simple relation:

$$\varepsilon_0\mu_0c^2 = 1 \qquad (3.2.30)$$

implying once again that the vacuum in classical electrodynamics has an electromagnetic essence.

$W_\omega(\vec{r},t)$ has to be understood as a time varying scalar field entailing that the energy density has not fixed values at any space coordinate, due naturally to the periodic variation of the electric and magnetic fields. Frequently, $W_\omega(\vec{r},t)$ is expressed through the vacuum permittivity only, getting for the total energy of a monochromatic electromagnetic field in a volume V:

$$E(\omega,\varepsilon_0) = \frac{\varepsilon_0}{2}\int_V (|\vec{E}_\omega(\vec{r},t)|^2 + c^2|\vec{B}_\omega(\vec{r},t)|^2)d^3r \qquad (3.2.31)$$

It is quite important noticing that the energy of the electromagnetic waves depends directly on the vacuum intrinsic properties, the electric permittivity ε_0 and magnetic permeability μ_0.

Furthermore, E_ω depends on the integration volume V and it is a time varying function.

• TEM waves — Laplace equation

General solutions of Maxwell's equations for TEM waves are without longitudinal field components.

From equations (3.1.5), (3.2.2) and (3.2.3), we can write the curls of the electric and magnetic field as following

$$\vec{\nabla} \times \vec{E}\,(\vec{r}, t) = -iw\,\mu\vec{H}\,(\vec{r}, t) \tag{3.2.32}$$

$$\vec{\nabla} \times \vec{H}\,(\vec{r}, t) = iw\,\varepsilon\vec{E}\,(\vec{r}, t) \tag{3.2.33}$$

Using the plane wave expression (3.2.22) and (3.2.23) and considering the propagation along the z-axis so that in Cartesian coordinates the components of the electric and magnetic field intensities of the transverse wave depend on (x, y), the equations (3.2.32) and (3.2.33) are expanded to the component expressions

$$\left[\frac{\partial E_z}{\partial y} + i\gamma E_y\right] + iw\mu H_x = 0 \tag{3.2.34}$$

$$\left[\frac{\partial E_z}{\partial x} + i\gamma E_x\right] - iw\mu H_y = 0 \tag{3.2.35}$$

$$\left[\frac{\partial E_y}{\partial x} - \frac{\partial E_x}{\partial y}\right] + iw\mu H_z = 0 \tag{3.2.36}$$

$$\left[\frac{\partial H_z}{\partial y} + i\gamma H_y\right] - iw\varepsilon\,E_x = 0 \tag{3.2.37}$$

$$\left[\frac{\partial H_z}{\partial x} + i\gamma H_x\right] + iw\varepsilon\,E_y = 0 \tag{3.2.38}$$

$$\left[\frac{\partial H_y}{\partial x} - \frac{\partial H_x}{\partial y}\right] - iw\varepsilon\,E_z = 0 \tag{3.2.39}$$

where we have introduced a general wave vector γ, called propagation factor, in such a way that the difference between the square values of k (given by 3.2.12) and γ define a cut-off wave vector k_c

$$k_c^2 = k^2 - \gamma^2 \tag{3.2.40}$$

According to the last set of six components equations, it is important noting that all the transverse field components can be calculated with respect to the longitudinal components E_z and H_z

$$E_x = -\frac{i}{k_c^2}\left[\gamma\frac{\partial E_z}{\partial x} + \omega\mu\frac{\partial H_z}{\partial y}\right] \qquad (3.2.41)$$

$$E_y = -\frac{i}{k_c^2}\left[\gamma\frac{\partial E_z}{\partial y} - \omega\mu\frac{\partial H_z}{\partial x}\right] \qquad (3.2.42)$$

$$H_x = \frac{i}{k_c^2}\left[\omega\varepsilon\frac{\partial E_z}{\partial y} - \gamma\frac{\partial H_z}{\partial x}\right] \qquad (3.2.43)$$

$$H_y = -\frac{i}{k_c^2}\left[\omega\varepsilon\frac{\partial E_z}{\partial x} + \gamma\frac{\partial H_z}{\partial y}\right] \qquad (3.2.44)$$

Hence, the components of TE waves, for which $E_z = 0$, as well as TM waves, for which $H_z = 0$, are immediately deduced from equations (3.2.41) to (3.2.44).

Now, one may wonder what happens for a TEM wave, when both E_z and H_z are zero. Obviously, we cannot use equations (3.2.41) to (3.2.44). Hence, we have to apply again equations (3.2.34) to (3.2.38) from which, by eliminating H_x, we get simply

$$\mu\varepsilon\omega^2 E_y = \gamma^2 E_y \rightarrow \gamma = k = \omega\sqrt{\mu\varepsilon} \qquad (3.2.45)$$

entailing that the cut-off wave vector is zero for TEM waves

$$k_c(TEM) = 0 \qquad (3.2.46)$$

From the Helmholtz equation, the (3.2.45) relation and the $e^{-i\gamma z}$ dependence of each component of the field we get

$$\left(\frac{\partial^2}{\partial x^2} + \frac{\partial^2}{\partial y^2} + \frac{\partial^2}{\partial z^2} + k^2\right)\begin{pmatrix} E_x \\ E_y \end{pmatrix} = \begin{pmatrix} 0 \\ 0 \end{pmatrix} \qquad (3.2.47)$$

Because $\frac{\partial^2}{\partial z^2}\begin{pmatrix} E_x \\ E_y \end{pmatrix} = -\gamma^2\begin{pmatrix} E_x \\ E_y \end{pmatrix} = -k^2\begin{pmatrix} E_x \\ E_y \end{pmatrix}$ equation (3.2.47) becomes the Laplace equation for the transverse components of a TEM wave

$$\left(\frac{\partial^2}{\partial x^2} + \frac{\partial^2}{\partial y^2}\right)\begin{pmatrix} E_x \\ E_y \end{pmatrix} = \begin{pmatrix} 0 \\ 0 \end{pmatrix} \qquad (3.2.48)$$

It is easy to demonstrate that the magnetic field transverse components H_x, H_y also satisfy Laplace equation, entailing that the transverse components of the TEM wave behave as static fields between conductors.

3.3. Scalar and Vector Potentials

The classical electromagnetic theory described briefly above represents light as electromagnetic plane waves whose electric and magnetic fields may oscillate at any frequency from "zero" to "infinity". From mathematical point of view, $\vec{E}_\omega(\vec{r},t)$ and $\vec{B}_\omega(\vec{r},t)$ can be generated from the coupling of scalar and vector potentials which are frequently denoted by $\Phi_\omega(\vec{r},t)$ and $\vec{A}_\omega(\vec{r},t)$ respectively

$$\vec{E}_\omega(\vec{r},t) = -\vec{\nabla}\,\Phi_\omega(\vec{r},t) - \frac{\partial}{\partial t}\vec{A}_\omega(\vec{r},t) \qquad (3.3.1)$$

$$\vec{B}_\omega(\vec{r},t) = \vec{\nabla} \times \vec{A}_\omega(\vec{r},t) \qquad (3.3.2)$$

Obviously, the scalar and vector potentials are not precisely defined by the last relations since equations (3.3.1) and (3.3.2) remain unaltered when considering any function $\sigma(\vec{r},t)$ such as

$$\Phi_\omega(\vec{r},t) \rightarrow \Phi_\omega(\vec{r},t) - \frac{\partial}{\partial t}\sigma(\vec{r},t) \qquad (3.3.3)$$

and

$$\vec{A}_\omega(\vec{r},t) \rightarrow \vec{A}_\omega(\vec{r},t) + \vec{\nabla}\sigma(\vec{r},t) \qquad (3.3.4)$$

In electromagnetic theory, this property is called Gauge Invariance which requires a supplementary condition for the definition of the scalar and vector potential $\Phi_\omega(\vec{r},t)$ and $\vec{A}_\omega(\vec{r},t)$.

In Lorentz Gauge, the supplementary condition is chosen to be

$$\vec{\nabla} \cdot \vec{A}_\omega(\vec{r},t) = -\mu\varepsilon\frac{\partial}{\partial t}\Phi_\omega(\vec{r},t) \qquad (3.3.5)$$

From the propagation equation (3.2.20) and (3.3.1), in absence of charges and current densities $|\vec{j}| = 0$, $\rho = 0$, and using the Lorentz Gauge condition, we get

$$\vec{\nabla}^2 \vec{A}_\omega(\vec{r},t) - \mu\varepsilon\frac{\partial^2}{\partial t^2}\vec{A}_\omega(\vec{r},t) = 0 \qquad (3.3.6)$$

and

$$\vec{\nabla}^2 \Phi_\omega(\vec{r}, t) - \mu\varepsilon \frac{\partial^2}{\partial t^2} \Phi_\omega(\vec{r}, t) = 0 \qquad (3.3.7)$$

showing that all the electric and magnetic fields of a monochromatic electromagnetic wave, $\vec{E}_\omega(\vec{r}, t)$, $\vec{B}_\omega(\vec{r}, t)$, $\vec{A}_\omega(\vec{r}, t)$ and $\Phi_\omega(\vec{r}, t)$ propagate in space satisfying, exactly the same propagation equation.

If we consider that the current and charge densities are not zero, it is easy to get from Maxwell's equations (3.1.8) to (3.1.11) the complete propagation equations:

$$\vec{\nabla}^2 \vec{A}_\omega(\vec{r}, t) - \mu\varepsilon \frac{\partial^2}{\partial t^2} \vec{A}_\omega(\vec{r}, t) = -\mu \vec{J}(\vec{r}, t) \qquad (3.3.8)$$

and

$$\vec{\nabla}^2 \Phi_\omega(\vec{r}, t) - \mu\varepsilon \frac{\partial^2}{\partial t^2} \Phi_\omega(\vec{r}, t) = -\frac{\rho(\vec{r}, t)}{\varepsilon} \qquad (3.3.9)$$

The general solutions of the last equations are:

$$\vec{A}(\vec{r}, t) = \frac{\mu}{4\pi} \int \frac{\vec{J}\left(\vec{r}', t - \frac{|\vec{r} - \vec{r}'|}{c}\right)}{|\vec{r} - \vec{r}'|} d^3r \qquad (3.3.10)$$

and

$$\Phi(\vec{r}, t) = \frac{1}{4\pi\varepsilon} \int \frac{\rho\left(\vec{r}', t - \frac{|\vec{r} - \vec{r}'|}{c}\right)}{|\vec{r} - \vec{r}'|} d^3r \qquad (3.3.11)$$

in which the time delay appearing in the current and charge densities is conditioned by the speed of the electromagnetic waves in vacuum, c.

At that level it is very important to note that the dimension analysis of (3.3.10) shows that the vector potential is inversely proportional to time, hence proportional to a frequency.

Now, let us consider a different Gauge condition which considerably simplifies the equations. In the so called Coulomb Gauge, the supplementary condition is chosen as:

$$\vec{\nabla} \cdot \vec{A}_\omega(\vec{r}, t) = 0 \qquad (3.3.12)$$

From (3.3.5) and (3.3.9), we get Poisson's equation relating the scalar potential to the charge density:

$$\vec{\nabla}^2 \Phi_\omega(\vec{r}, t) = -\frac{\rho(\vec{r}, t)}{\varepsilon} \tag{3.3.13}$$

whose solution is now independent of the time delay characterizing the charge density:

$$\Phi(\vec{r}, t) = \frac{1}{4\pi\varepsilon} \int \frac{\rho(\vec{r}', t')}{|\vec{r} - \vec{r}'|} d^3 r' \tag{3.3.14}$$

Hence in the absence of charges $\Phi(\vec{r}, t) = 0$ and from (3.3.1), we get a quite important relation between the electric field and the vector potential in the Coulomb Gauge.

$$\vec{E}_\omega(\vec{r}, t) = -\frac{\partial}{\partial t} \vec{A}_\omega(\vec{r}, t) \tag{3.3.15}$$

The last relation is principally used in classical and quantum electrodynamics to calculate the electric field starting from the vector potential.

3.4. Vector Potential and Electromagnetic Field Polarization

Using the above relations, the propagation equation for the vector potential in vacuum writes

$$\vec{\nabla}^2 \vec{A}_\omega(\vec{r}, t) - \frac{1}{c^2} \frac{\partial^2}{\partial t^2} \vec{A}_\omega(\vec{r}, t) = 0 \tag{3.4.1}$$

In the case of a plane monochromatic electromagnetic field the vector potential is real (from mathematical point of view) and it is expressed in the same way as the electric and magnetic fields

$$\vec{A}_\omega(\vec{r}, t) = \vec{A}_0(\omega) \left(e^{-i(\vec{k}\cdot\vec{r} - \omega t + \theta)} + c.c. \right)$$
$$= 2\vec{A}_0(\omega) \cos(\vec{k}\cdot\vec{r} - \omega t + \theta) \tag{3.4.2}$$

where θ is a phase parameter.

For the Coulomb Gauge condition to be satisfied, the following relation between the wave vector and the amplitude of the vector potential should hold:

$$\vec{k} \cdot \vec{A}_0(\omega) = 0 \tag{3.4.3}$$

entailing that the wave vector is perpendicular to the vector potential so that the electromagnetic wave is called transverse.

Taking $\theta = 0$ and using the relations (3.3.15) and (3.3.2), while putting $\vec{A}_0 \rightarrow A_0(\omega)\,\hat{\varepsilon}$ and $\vec{k} = \frac{\omega}{c}\hat{k}$ we get directly the electric field and magnetic induction expressed through the vector potential

$$\vec{E}_\omega(\vec{r},t) = -2\omega\,A_0(\omega)\,\hat{\varepsilon}\sin\left(\vec{k}\cdot\vec{r}-\omega\,t\right) \tag{3.4.4}$$

$$\vec{B}_\omega(\vec{r},t) = -\frac{1}{c}2\,\omega A_0(\omega)(\hat{k}\times\hat{\varepsilon})\sin\left(\vec{k}\cdot\vec{r}-\omega\,t\right) \tag{3.4.5}$$

The polarization of the radiation, i.e., the spatial orientation of the electric field, is also specified by the unit vector of the vector potential $\hat{\varepsilon}$.

From the coupling of the equation (3.2.28) with the last expressions one gets in vacuum,

$$\langle\vec{S}_\omega\rangle_T = 2\varepsilon_0 c\,\omega^2 A_0^2(\omega)\,\hat{k} \tag{3.4.6}$$

expressing that the power density along the propagation axis (Poynting vector) is also expressed through the vector potential over a period T.

In the same way the energy density of the electromagnetic field is obtained by (3.2.29), (3.4.4) and (3.4.5):

$$W_\omega(\vec{r},t) = 4\varepsilon_0\omega^2 A_0^2(\omega)\sin^2(\vec{k}\cdot\vec{r}-\omega t) \tag{3.4.7}$$

whose mean value over a period writes:

$$\langle W_\omega\rangle_T = 2\varepsilon_0\omega^2 A_0^2(\omega) = \frac{1}{c}|\langle\vec{S}_\omega\rangle_T| \tag{3.4.8}$$

Consequently, the only knowledge of the vector potential amplitude $A_0(\omega)$ is sufficient to deduce the principal physical properties of a TMEM wave of angular frequency ω in vacuum.

We can now have a close look to the vector orientation of the electric field and that of the vector potential with respect to the propagation axis. Following the equation (3.4.3), the unit vector of the vector potential and thus that of the electric field intensity, $\hat{\varepsilon}$, lies in a plane which is perpendicular to the propagation vector $\hat{k}$.

Considering two orthogonal unit vectors $\hat{e}_1$ and $\hat{e}_2$, defining the plane perpendicular to the wave vector $\hat{k}$ forming a right-handed system:

$$\hat{k} = \hat{e}_1 \times \hat{e}_2, \quad \hat{e}_1 \cdot \hat{e}_2 = 0 \qquad (3.4.9)$$

so as the unit vector $\hat{\varepsilon}$ of the vector potential can be decomposed simply as

$$\hat{\varepsilon} = \alpha_1 \hat{e}_1 + \alpha_2 \hat{e}_2 \qquad (3.4.10)$$

And since $\hat{\varepsilon}$ is a unit vector, the following relation between the coefficients α_1 and α_2 holds

$$\alpha_1^2 + \alpha_2^2 = 1 \qquad (3.4.11)$$

Thus, we can define two complex orthogonal unit vectors $\hat{L}$ and $\hat{R}$ such as

$$\hat{L} = \frac{1}{\sqrt{2}}(\hat{e}_1 + i\hat{e}_2) \quad \text{and} \quad \hat{R} = \frac{1}{\sqrt{2}}(\hat{e}_1 - i\hat{e}_2) \qquad (3.4.12)$$

appropriate for describing respectively, the Left-hand and Right-hand circularly polarized electromagnetic waves, corresponding to two states of the radiation polarization.

In the first case, the electric field follows a circular left-hand precession along the propagation axis with the angular frequency ω, while in the second one it follows a circular right-hand precession at the same angular frequency.

From equations (3.4.2) and (3.4.12), one may easily deduce the components of the vector potential of a circularly polarized plane wave in Cartesian coordinates ($\hat{e}_1 \rightarrow \hat{x}$ and $\hat{e}_2 \rightarrow \hat{y}$):

$$A_x^{(L)}(z,t) = A_x^{(R)}(z,t) = \sqrt{2}A_0(\omega) \cos(kz - \omega t)$$
$$A_y^{(L)}(z,t) = -A_y^{(R)}(z,t) = -\sqrt{2}A_0(\omega) \sin(kz - \omega t)$$
$$A_z^{(L)} = A_z^{(R)} = 0 \qquad (3.4.13)$$

Using (3.3.15) and (3.4.13), the expressions of the corresponding electric field components write:

$$E_x^{(L)}(z,t) = E_x^{(R)}(z,t) = \omega A_y^{(L)} = -\omega A_y^{(R)}$$

$$E_y^{(L)}(z,t) = -E_y^{(R)}(z,t) = -\omega A_x^{(L)} = -\omega\, A_x^{(R)}$$
$$E_z^{(L)} = E_z^{(R)} = 0 \tag{3.4.14}$$

Consequently, an arbitrary state of polarization can be obtained by the complex unit vector $\hat{P}$ consisting of a linear combination of $\hat{L}$ and $\hat{R}$:

$$\hat{P} = \beta_L\hat{L} + \beta_R\hat{R}$$
$$= \frac{1}{\sqrt{2}}(\beta_L(\hat{x} + i\hat{y}) + \beta_R(\hat{x} - i\hat{y})) \quad \text{with } |\beta_L|^2 + |\beta_R|^2 = 1$$

$$\tag{3.4.15}$$

For example, a linear polarization consists of the oscillation of the electric and magnetic field in the orthogonal planes (x, z) and (y, z) along the propagation axis z.

Obviously, such a polarization is easily obtained by the combination of two circularly polarized waves propagating in phase and having right and left polarization.

3.5. Guided Propagation of Electromagnetic Waves

In the previous chapter, we have seen that the electric and magnetic fields of a monochromatic wave obey the propagation equation (3.2.26) in free space. It is of high interest to explore the way the electromagnetic wave propagate in a waveguide of specific cross sectional shape and dimensions. The problem consists of seeking solutions to Maxwell's equations satisfying the particular boundary conditions imposed by the waveguide shape.

Let us consider the simple case in which the inner walls of the waveguide consists of a homogeneous and isotropic dielectric with fixed values of ε and μ. It is worth noting that a TEM mode can only propagate in the presence of two or more guiding conductors. Consequently, a TEM mode cannot propagate in a hollow waveguide since it is composed of a single conductor. Thus, we give below the components of the electric and magnetic fields of transverse electric (TE) and transverse magnetic (TM) modes satisfying Maxwell's equations and the boundary conditions, in the case of rectangular and circular waveguides.

- Rectangular waveguide of cross section a, b $(a > b)$

We also assume here time-harmonic fields with an angular frequency ω, hence with an exponential dependence $e^{i\omega t}$. The propagating fields must be solutions of Maxwell's equations satisfying the boundary conditions, for the components of the electric and magnetic fields on the waveguide walls.

3.5.1. $TE_{m,n}$ Modes

Are characterized by the absence of the electric field component along the propagation axis $E_z = 0$. The magnetic field components H_z has to satisfy the reduced (simplifies the time dependence) propagation equation

$$\left[\frac{\partial^2}{\partial x^2} + \frac{\partial^2}{\partial y^2} + k_c^2 \right] H_z(x, y) = 0 \qquad (3.5.1)$$

The boundary conditions for the electric field tangential components are

$$\hat{e}_x(x, y) = 0 \quad \text{at } (y = 0 \text{ and } b) \quad \text{and}$$
$$\hat{e}_y(x, y) = 0 \quad \text{at } (x = 0 \text{ and } a) \qquad (3.5.2)$$

The solutions in Cartesian coordinates (x, y, z) are given by

$$H_x = iH_{0_z} \frac{\gamma}{k_c^2} \left(\frac{m\pi}{a} \right) \sin \left(\frac{m\pi x}{a} \right) \cos \left(\frac{n\pi y}{b} \right) e^{-i\gamma z} \qquad (3.5.3)$$

$$H_y = iH_{0_z} \frac{\gamma}{k_c^2} \left(\frac{n\pi}{b} \right) \cos \left(\frac{m\pi x}{a} \right) \sin \left(\frac{n\pi y}{b} \right) e^{-i\gamma z} \qquad (3.5.4)$$

$$H_z = H_{0_z} \cos \left(\frac{m\pi x}{a} \right) \cos \left(\frac{n\pi y}{b} \right) e^{-i\gamma z} \qquad (3.5.5)$$

$$E_x = H_y \left(\frac{\omega\mu}{\gamma} \right); \quad E_y = -H_x \left(\frac{\omega\mu}{\gamma} \right); \quad E_z = 0 \qquad (3.5.6)$$

Where γ is the propagation factor

$$\gamma = \sqrt{k^2 - k_c^2} = \left(\mu\varepsilon\omega^2 - \left(\frac{m\pi}{a} \right)^2 - \left(\frac{n\pi}{b} \right)^2 \right)^{1/2} \qquad (3.5.7)$$

with $k^2 = \mu\varepsilon\omega^2$ and

$$k_c^2 = \left(\frac{2\pi}{\lambda_c}\right)^2 = \left(\frac{m\pi}{a}\right)^2 + \left(\frac{n\pi}{b}\right)^2 \qquad (3.5.8)$$

is the cut-off wave vector and λ_c the cut-off wavelength corresponding to a cut-off angular frequency

$$\omega_c = \frac{k_c}{\sqrt{\mu\varepsilon}} \quad \text{and} \quad \lambda_c = \frac{2\pi}{\sqrt{\left(\frac{m\pi}{a}\right)^2 + \left(\frac{n\pi}{b}\right)^2}} \qquad (3.5.9)$$

Only modes with $\omega > \omega_c$ and $\lambda < \lambda_c$ can propagate in this rectangular waveguide.

3.5.2. *TM$_{m,n}$ Components*

The $TM_{m,n}$ components are characterized by the absence of the magnetic field component along the propagation axis $H_z = 0$. The electric field components E_z has to satisfy the reduced propagation equation

$$\left[\frac{\partial^2}{\partial x^2} + \frac{\partial^2}{\partial y^2} + k_c^2\right] E_z(x,y) = 0 \qquad (3.5.10)$$

The boundary conditions for $\hat{e}_x$ and $\hat{e}_y$ are also satisfied by $\hat{e}_z$, i.e.,

$$\hat{e}_z(x,y) = 0 \quad \text{at } (x = 0 \text{ and } a) \quad \text{and} \quad \hat{e}_z(x,y) = 0 \quad \text{at } (y = 0 \text{ and } b) \qquad (3.5.11)$$

The solutions in Cartesian coordinates (x, y, z) are given by

$$E_x = -iE_{0_z}\frac{\gamma}{k_c^2}\left(\frac{m\pi}{a}\right)\cos\left(\frac{m\pi x}{a}\right)\sin\left(\frac{n\pi y}{b}\right)e^{-i\gamma z} \qquad (3.5.12)$$

$$E_y = -iE_{0_z}\frac{\gamma}{k_c^2}\left(\frac{n\pi}{b}\right)\sin\left(\frac{m\pi x}{a}\right)\cos\left(\frac{n\pi y}{b}\right)e^{-i\gamma z} \qquad (3.5.13)$$

$$E_z = E_{0_z}\sin\left(\frac{m\pi x}{a}\right)\sin\left(\frac{n\pi y}{b}\right)e^{-i\gamma z} \qquad (3.5.14)$$

$$H_x = -E_y\left(\frac{\omega\varepsilon}{\gamma}\right); \quad H_y = E_x\left(\frac{\omega\varepsilon}{\gamma}\right); \quad H_z = 0 \qquad (3.5.15)$$

where n, m are integers (including 0). H_{0_z} and E_{0_z} are the amplitudes of the magnetic and the electric field respectively along the direction of propagation z.

As in the case of $TE_{m,n}$ modes in a rectangular waveguide, the cut-off wavelength λ_c for the $TM_{m,n}$ writes

$$\lambda_c = \frac{2\pi}{\sqrt{\left(\frac{m\pi}{a}\right)^2 + \left(\frac{n\pi}{b}\right)^2}} \qquad (3.5.16)$$

Hence, in a rectangular waveguide, only $TE_{m,n}$ and $TM_{m,n}$ modes with a wavelength shorter than λ_c can propagate.

- Circular waveguide of radius r_0

In a circular waveguide things are slightly different and it is more practical to use the cylindrical coordinates (r, θ, z).

3.5.3. $TE_{m,n}$ Modes

In the case of the rectangular waveguide, the electric field component along the propagation axis $E_z = 0$, while the magnetic field component H_z is a solution to the equation

$$\left[\frac{\partial^2}{\partial\rho^2} + \frac{1}{\rho}\frac{\partial}{\partial\rho} + \frac{1}{\rho^2}\frac{\partial^2}{\partial\theta^2} + k_c^2\right] H_z(\rho, \theta) = 0 \qquad (3.5.17)$$

The solution for $H_z(\rho, \theta)$ has to be periodic in angular rotation so that $H_z(\rho, \theta) = H_z(\rho, \theta + 2n\pi)$, while the tangential component of the electric field on the wall should vanish as $E_\theta(\rho = r_0, \theta) = 0$.

Respecting these boundary conditions the field components have the expressions

$$H_z = H_{0_z} \cos(m\theta) J_m\left(R_n^{(J_m')}\frac{\rho}{r_0}\right) e^{-i\gamma z} \qquad (3.5.18)$$

$$H_\theta = H_{0_z}\frac{\gamma}{k_c^2}\left(\frac{m}{\rho}\right)\sin(m\theta) J_m\left(R_n^{(J_m')}\frac{\rho}{r_0}\right) e^{-i\gamma z} \qquad (3.5.19)$$

$$H_\rho = -iH_{0_z}\left(\frac{\gamma}{k_c}\right)\cos(m\theta) J_m'\left(R_n^{(J_m')}\frac{\rho}{r_0}\right) e^{-i\gamma z} \qquad (3.5.20)$$

$$E_\theta = -H_\rho\left(\frac{\omega\mu}{\gamma}\right); \quad E_\rho = H_\theta\left(\frac{\omega\mu}{\gamma}\right); \quad E_z = 0 \qquad (3.5.21)$$

where J_m is the Bessel function of the first kind of order m and $R_n^{(J'_m)}$ is the nth root of J'_m, the derivative with respect to the argument. The cut-off wavelength λ_c and the corresponding cut-off angular frequency ω_c are also given as a function of the circular waveguide cross section parameter r_0.

$$\lambda_c = \frac{2\pi \, r_0}{R_n^{(J'_m)}} \tag{3.5.22}$$

$$\omega_c = \frac{R_n^{(J'_m)}}{r_0\sqrt{\mu\varepsilon}} \tag{3.5.23}$$

In a circular waveguide with radius r_0 only the $TE_{m,n}$ modes with a wavelength shorter than λ_c can propagate.

3.5.4. $TM_{m,n}$ Modes

The TM modes are obtained by solving for E_z, the propagation equation:

$$\left[\frac{\partial^2}{\partial\rho^2} + \frac{1}{\rho}\frac{\partial}{\partial\rho} + \frac{1}{\rho^2}\frac{\partial^2}{\partial\theta^2} + k_c^2 \right] E_z(\rho, \theta) = 0 \tag{3.5.24}$$

This is an identical equation to that of the H_z component in the TE case. Thus, the general solutions have the same form and are obtained with the boundary condition that the z component of the electric field vanishes at the walls: $E_z(\rho = r_0, \theta) = 0$.

$$E_z = E_{0_z} \cos(m\theta) J_m\left(R_n^{(J_m)}\frac{\rho}{r_0}\right) e^{-i\gamma z} \tag{3.5.25}$$

$$E_\theta = iE_{0_z}\frac{\gamma}{k_c^2}\left(\frac{m}{\rho}\right) \sin(m\theta) J_m\left(R_n^{(J_m)}\frac{\rho}{r_0}\right) e^{-i\gamma z} \tag{3.5.26}$$

$$E_\rho = -iE_{0_z}\left(\frac{\gamma}{k_c}\right) \cos(m\theta) J'_m\left(R_n^{(J_m)}\frac{\rho}{r_0}\right) e^{-i\gamma z} \tag{3.5.27}$$

$$H_\theta = E_\rho\left(\frac{\omega\varepsilon}{\gamma}\right); \quad H_\rho = -E_\theta\left(\frac{\omega\varepsilon}{\gamma}\right); \quad H_z = 0 \tag{3.5.28}$$

The expression for the propagation factor is the same as in the TE modes while the cut-off wavelength and angular frequency cut-off now write

$$\lambda_c = \frac{2\pi \, r_0}{R_n^{(J_m)}} \qquad (3.5.29)$$

and

$$\omega_c = \frac{R_n^{(J_m)}}{r_0 \sqrt{\mu\varepsilon}} \qquad (3.5.30)$$

where $R_n^{(J_m)}$ is the nth root of J_m, that is $J_m(R_n^{(J_m)}) = 0$.

- Case of two parallel plates

This is a particular configuration that accepts TE and TM waves and can also support TEM waves since it is composed of two conductors. As we have seen in 3.2, the TEM waves behave as the static fields between conductors and consequently they can be deduced from the scalar potential that satisfies Poisson's equation (3.3.13) in the plane (x, y) and in the absence of charges.

$$\vec{\nabla}_{(x,y)}^2 \, \Phi(x, y) = 0 \qquad (3.5.31)$$

Considering two parallel plates in the (x, z) plane separated by a distance y_0, having a potential difference of U, and considering a definite length along the x axis, x_0, for $0 \le x \le x_0$ and $0 \le y \le y_0$, the boundary conditions for the scalar potential write

$$\Phi(x, 0) = 0 \quad \text{and} \quad \Phi(x, y_0) = U \qquad (3.5.32)$$

The solution of (3.5.31) respecting the boundary conditions (3.5.32) is

$$\Phi(x, y) = \frac{y}{y_0} U \qquad (3.5.33)$$

The TE field is obtained immediately

$$\vec{E}(x, y) = -\vec{\nabla}_{(x,y)} \Phi(x, y) = \frac{1}{y_0} U \, \hat{e}_y \qquad (3.5.34)$$

and the complete expression of the electric field is

$$\vec{E}(x, y, z) = \frac{U}{y_0} e^{-ikz} \, \hat{e}_y \qquad (3.5.35)$$

and the magnetic field becomes

$$\vec{H}(x, y, z) = \frac{U}{y_0 R} e^{-ikz} \, \hat{e}_x \qquad (3.5.36)$$

where R is the impendence of the medium between the parallel plates given by (3.2.14). Since E_z and H_z are zero here, the TEM fields in this situation are analogue to plane waves in a homogeneous medium.

Now, one can easily calculate the potential $V(y_0)$ on the plate at $y = y_0$ from the classical electrostatic formalism

$$V(y_0) = -\int_0^{y_0} \frac{U}{y_0} e^{-ikz}\, dy = U\, e^{-ikz} \qquad (3.5.37)$$

while the surface current density on the same plate is calculated as following

$$I(y_0) = -\int_0^{x_0} (\hat{e}_y \times \vec{H}) \cdot \hat{e}_z\, dx = \frac{x_0}{y_0} \frac{U}{R} e^{-ikz} \qquad (3.5.38)$$

This is a quite interesting result showing that TEM waves can propagate between two parallel plates with a potential difference U, inducing a current density on the upper plate.

• Density of states

The guided propagation of the electromagnetic waves reveals that whatever the shape of the box the exponential part of the wave function $e^{-i\vec{k}\cdot\vec{r}}$ of (3.4.2) suffers the boundary conditions. This introduces naturally the notion that the electromagnetic wave is composed of "states", whose number is imposed by the shape and the dimensions of a given volume and can be calculated.

Let us consider the simple case of a cube of a large side L and volume V. Then the periodic boundary conditions, like in (3.5.8) for the wave vector are

$$k_x = \frac{2\pi}{L} n_x, \quad k_y = \frac{2\pi}{L} n_y, \quad k_z = \frac{2\pi}{L} n_z \qquad (3.5.39)$$

with $n_x, n_y, n_z = 0, \pm 1, \pm 2, \ldots$.

The number of modes in the range $dk_x dk_y dk_z = d^3 k$ writes

$$dk_x dk_y dk_z = \left(\frac{2\pi}{L}\right)^3 dn_x dn_y dn_z \qquad (3.5.40)$$

Assuming that L is too large compared to the wavelength, n_x, n_y, n_z can be considered as continuous variables

$$dn_x dn_y dn_z = \left(\frac{L}{2\pi}\right)^3 dk_x dk_y dk_z = \frac{V}{8\pi^3} 4\pi k^2 dk \qquad (3.5.41)$$

Using the dispersion relation $\omega = kc$ and considering two directions of polarization per state, the total number of states in the angular frequency interval $d\omega$ writes

$$dn(\omega) = V \frac{\omega^2}{\pi^2 c^3} d\omega \qquad (3.5.42)$$

Notice that the same expressions will be used later in quantum mechanics for the calculation of the number of photons in a volume V, revealing an interesting relation between the notion of electromagnetic field "state" and the concept of the "photon".

Hence, the states density, corresponding to the number of states per unit volume and unit frequency $\nu = \omega/2\pi$ results readily from (3.5.42),

$$\rho(\nu) = \frac{8\pi\nu^2}{c^3} \qquad (3.5.43)$$

Equation (3.5.41) is extremely useful for transforming the discrete summation over the wave vector k, characterizing each state to a continuous summation over the angular frequencies

$$\sum_k \rightarrow \frac{V}{8\pi^3} \int 4\pi k^2 dk = \frac{V}{2\pi^2 c^3} \int \omega^2 d\omega \qquad (3.5.44)$$

This is the key relation used in all calculations in quantum electrodynamics in order to obtain physical results independent of the volume parameter V.

3.6. Conclusion Remarks

The principal conclusion that can be drawn out from the last chapter is that the electromagnetic waves can propagate in vacuum and the propagation equations are immediately deduced from Maxwell's equations.

However, in Maxwell's theory the vacuum itself is not a completely empty medium but has an electromagnetic nature, processing a magnetic permeability $\mu_0 = 4\pi \times 10^{-7} \mathrm{H\,m^{-1}}$ and an electric permittivity $\varepsilon_0 = 8.85 \times 10^{-12} \mathrm{F\,m^{-1}} \approx \frac{1}{36\pi} 10^{-9} \mathrm{F\,m^{-1}}$, and consequently an impendence of a precise value, $R_{\mathrm{vacuum}} = \sqrt{\frac{\mu_0}{\varepsilon_0}} \approx 120\pi\,\Omega$.

Whereas, the speed c of the electromagnetic waves is imposed by the intrinsic electric and magnetic properties of the vacuum, $c = \frac{1}{\sqrt{\varepsilon_0\mu_0}} \approx 2.9979\,10^8\,\mathrm{m\,s^{-1}}$, which at the present status of knowledge is considered to be the highest speed permitted in the universe.

Furthermore, it is important noticing that the energy density and flux of the electromagnetic waves depend equally on the vacuum properties ε_0 and μ_0.

It turns out that the electromagnetic waves, in other words light, appear to behave as a natural perturbation of the vacuum, having intrinsic electric and magnetic properties. The vector potential of the electromagnetic field, issued from Maxwell's equations is proportional to the angular frequency.

On the other hand, it is of high interest to notice the volume importance for the existence of a "mode" of the electromagnetic field, issued from the guided propagation studies. All the cut-off wavelengths calculated for the different shapes of waveguides entail that longer wavelengths beyond those values cannot stand within a volume of a given shape and dimensions. Consequently, an electromagnetic wave mode of a given wavelength, even the simplest one composed by a single state, cannot subsist in a volume whose dimensions correspond to a smaller cut-off wavelength.

Finally, the number of states per unit volume and unit frequency calculated by $\rho(\nu) = \frac{8\pi\nu^2}{c^3}$ is only valid in a volume whose dimensions are considerably bigger than the wavelength of the electromagnetic waves.

Bibliography

1. C.L. Andrews, *Optics of the electromagnetic spectrum*, Englewood cliffs, New Jersey: Prentice-Hall Inc. 1960.
2. M. Born and E. Wolf, *Principles of optics*, Cambridge: Harvard University Press. 1999.
3. B.H. Bransden and C.J. Joachain, *Physics of atoms and molecules*, London: Longman Group Ltd. 1983.

4. S.L. Chuang, *Physics of photonic devices*, New Jersey: John Wiley & Sons. 2009.
5. J.C. Garrison and R.Y. Chiao, *Quantum optics*, Oxford: Oxford University Press. 2008.
6. H. Haken, *Light*, Amsterdam–Oxford: North Holland Publishing. 1981.
7. M.D. Pozar, *Microwave engineering*, 3rd Edition, New York: John Wiley & Sons, Inc. 2005.
8. F.H. Read, *Electromagnetic radiation*, New York: John Wiley & Sons. 1980.
9. B.E.A. Saleh and M.C. Teich, *Fundamentals of photonics*, New York: John Wiley & Sons. 2007.

Chapter 4

From Electromagnetic Waves to Quantum Electrodynamics

4.1. Elements of Quantum Mechanics

- Blackbody radiation and the ultraviolet catastrophe

The study of the properties of radiation emitted by hot bodies was the starting point for the development of the quantum theory at the end of the 19th and the beginning of the 20th centuries. The experimental evidence has shown that the spectral distribution, in other words the distribution of the wavelengths of the radiation, as well as the total energy radiated depend directly on the temperature of the body.

The number of standing electromagnetic waves, i.e., the number of modes, per unit volume and unit frequency within a cavity is given by the expression (3.5.43). It can be shown that for a sufficiently big cavity this expression holds whatever its shape. Considering, the electromagnetic waves to be in thermal equilibrium at temperature T with the walls of the cavity, the probability $p(E)$ that a mode has an energy E included between $E + dE$, is obtained in classical theory by

$$p(E) = \frac{e^{-E/kT}}{kT} \tag{4.1.1}$$

where k is Boltzmann's constant.

The mean energy of a mode is thus

$$\bar{E} = \int_0^\infty E p\left(E\right) dE = \int_0^\infty E \frac{e^{-E/kT}}{kT} dE = kT \qquad (4.1.2)$$

Consequently, all the modes have the same mean energy, independent of the frequency. Using (3.5.43), the energy per unit volume of the cavity at temperature T and per unit frequency writes

$$u\left(T, \nu\right) = \frac{8\pi\nu^2}{c^3} kT \qquad (4.1.3)$$

The last expression is known historically as the Rayleigh-Jeans energy density which fails to describe the experimentally observed results at high frequencies, often called the "ultraviolet catastrophe", while giving an infinite total energy per unit volume of the cavity when summing up along all the frequencies

$$U(T) = \int_0^\infty \frac{8\pi\nu^2}{c^3} kT d\nu \to \infty \qquad (4.1.4)$$

In order to resolve this problem, Max Planck introduced the original idea that the cavity walls are composed of local oscillators, each one being capable of absorbing and emitting discrete quantities of electromagnetic energy $n h\nu$, where n is an integer and h is a universal constant attributed in his honor to him, Planck's constant. This absolutely revolutionary concept at the beginning of the 20th century, led in the following years to the development of the quantum theory of the electronic energy levels in matter and in the quantization of the electromagnetic field.

Following Planck's idea, the total energy of a mode is $E_n = n h\nu$ and the probability for a mode to be composed of n quantum of electromagnetic energy, called photons, is

$$p_n = \gamma e^{-nh\nu/kT} \qquad (4.1.5)$$

where γ is a normalization factor so that the sum of all probabilities be equal to 1.

Hence,

$$\gamma \sum_n e^{-nh\nu/kT} = 1 \Rightarrow \gamma = 1 - e^{-h\nu/kT} \qquad (4.1.6)$$

Now, using $\beta = \frac{1}{kT}$, the mean energy of each mode in thermal equilibrium writes

$$\bar{\varepsilon} = \frac{\sum_0^\infty nh\nu\, e^{-nh\nu\beta}}{\sum_0^\infty e^{-nh\nu\beta}} = -\frac{d}{d\beta}\log\left(\sum_0^\infty e^{-nh\nu\beta}\right)$$

$$= -\frac{d}{d\beta}\log\left(\frac{1}{1-e^{-h\nu\beta}}\right) = \frac{h\nu}{e^{h\nu\beta}-1} \tag{4.1.7}$$

and the electromagnetic energy per unit volume and unit frequency now becomes within Planck's hypothesis

$$u\left(T,\nu\right) = \frac{8\pi\nu^2}{c^3}\frac{h\nu}{e^{h\nu/kT}-1} \tag{4.1.8}$$

The last expression is Planck's radiation law and interprets quite satisfactorily all the experimental results at any frequency and temperature of a black body.

- Energy and momentum operators in quantum mechanics

In the first two decades of the 20th century, the introduction by Max Planck of the quantized energy levels of the atoms and molecules resulting in quantized emissions and absorptions of the electromagnetic field, as well as many experiments demonstrating the wave properties of electrons (Davisson and Germer), created the necessity to elaborate a new theory.

The birth of quantum mechanics came along with Niels Bohr's and Erwin Schrödinger's works in 1925, based on the ideas put forward by Louis de Broglie in 1924, following which the energy and the momentum of particles are associated to a frequency and a wavelength, through the formalism already established previously for light

$$E = h\nu = \hbar\omega \quad \vec{P} = \frac{h}{\lambda}\hat{e}_k = \hbar\vec{k} \tag{4.1.9}$$

A plane wave is associated to the particle

$$\Psi\left(\vec{r},t\right) = \Psi_0\, e^{i(\vec{k}\cdot\vec{r}-\omega t)} \tag{4.1.10}$$

From the last two equations, one can easily deduce

$$i\hbar\frac{\partial}{\partial t}\Psi = E\Psi \quad -i\hbar\vec{\nabla}\Psi = \vec{p}\,\Psi \tag{4.1.11}$$

According to the last relations, a fundamental postulate of quantum mechanics was born according to which for a nonrelativistic particle, either it is free or not, the energy and the momentum are represented by the operators

$$E \rightarrow i\hbar\frac{\partial}{\partial t} \quad \vec{p} \rightarrow -i\hbar\vec{\nabla} \qquad (4.1.12)$$

At that point, we recall the correspondence already deduced from Maxwell's equations in (3.2) between the angular frequency and the time derivative operator $\omega \rightarrow i\frac{\partial}{\partial t}$ getting from (4.1.12), $E \rightarrow \hbar\omega$, when considering plane wave solutions.

Taking into account the potential energy V of the particle in the total energy and using (4.1.11), we get the Schrödinger equation which consists the basis of quantum mechanics

$$i\hbar\frac{\partial}{\partial t}\Psi = \left(\frac{\vec{P}^2}{2m} + V\right)\Psi = -\frac{\hbar^2}{2m}\vec{\nabla}^2\Psi + V\Psi \qquad (4.1.13)$$

To complement the fundamentals of the quantum mechanical theory, we can now consider the action of the momentum operator $p_\alpha = -i\hbar\frac{\partial}{\partial\alpha}$ along a Cartesian coordinate α where α is x, y or z acting on a function $\phi(\alpha)$. For that we apply the commutation operator $[\alpha, p_a] = \alpha p_\alpha - p_\alpha\alpha$ on a function $\phi(\alpha)$

$$[\alpha, p_a]\phi(\alpha) = -i\hbar\left(\alpha\frac{\partial\phi(\alpha)}{\partial\alpha} - \frac{\partial}{\partial\alpha}\alpha\phi(\alpha)\right) = i\hbar\phi(\alpha) \qquad (4.1.14)$$

The last result shows that the commutation between the coordinate α and the corresponding component of momentum operator p_α do not commute and can be generalized using the Cartesian coordinates

$$[x, p_x] = [y, p_y] = [z, p_z] = i\hbar \qquad (4.1.15)$$

All the other operators vanish, that is $[\alpha, p_{\alpha'}] = i\hbar\delta_{\alpha,\alpha'}$ where $\delta_{\alpha,\alpha'}$ is Kronecker symbol such as $\delta_{\alpha,\alpha'} = 1$ when $\alpha = \alpha'$ and $\delta_{\alpha,\alpha'} = 0$ for $\alpha \neq \alpha'$.

The physical consequence of this is that for the non-commuting operators a "simultaneous" measurement of the corresponding observables is impossible.

- Particle in a square potential well — correspondence with the waveguides

The simplest case of the application of Schrödinger equation is to consider a particle of mass m in a one-dimensional infinite square potential well, $V(z)$ defined by

$$V(z) = 0 \quad \text{for } |z| > z_0$$
$$V(z) \to \infty \quad \text{for } z < -z_0 \quad \text{and} \quad z > z_0 \qquad (4.1.16)$$

Since the potential tends to infinity for $z < -z_0$ and $z > z_0$, the particle has no possibility to be outside the borders of the potential, consequently its wave function $\Psi(z)$ must tend to zero in that interval.

For $|z| < z_0$ where $V(z) = 0$, the Schrödinger equation is simply written as

$$-\frac{\hbar^2}{2m} \frac{d^2}{dz^2} \Psi(z) = E \Psi(z) \qquad (4.1.17)$$

Putting

$$k = \frac{\sqrt{2mE}}{\hbar} \qquad (4.1.18)$$

the solution of (4.1.17) is a general trigonometric function of the form

$$\Psi(z) = c_1 \sin(kz) + c_2 \cos(kz) \qquad (4.1.19)$$

Taking into account that $\Psi(z)$ must be zero at $z = \pm z_0$ we get two solutions:
Either

$$c_1 = 0 \quad \text{and} \quad \cos(kz_0) = 0 \qquad (4.1.20)$$

so that the only permitted values of k are

$$k_n = \frac{n\pi}{2z_0} \quad (n = 1, 3, 5, 7, \ldots) \qquad (4.1.21)$$

Or

$$c_2 = 0 \quad \text{and} \quad \sin(kz_0) = 0 \qquad (4.1.22)$$

And the permitted values of k are now

$$k_n = \frac{n\pi}{2z_0} \quad (n = 2, 4, 6, 8, \ldots) \qquad (4.1.23)$$

Consequently the energy is deduced from (4.1.18) and is found to be quantified following the integer values of n

$$E_n = \frac{\hbar^2 k^2}{2m} = \frac{\hbar^2}{8m} \left(\frac{n\pi}{z_0} \right)^2 \quad (n = 1, 2, 3, 4, \ldots) \quad (4.1.24)$$

The wave function for both solutions can be normalized to unity by

$$\int_{-z_0}^{z_0} |\Psi_n(z)|^2 dz = 1 \quad (4.1.25)$$

getting for both solutions respectively

$$\Psi_n(z) = \frac{1}{z_0^{1/2}} \cos \left(\frac{n\pi}{2z_0} z \right) \quad (n = 1, 3, 5, \ldots) \quad (4.1.26)$$

and

$$\Psi_n(z) = \frac{1}{z_0^{1/2}} \sin \left(\frac{n\pi}{2z_0} z \right) \quad (n = 2, 4, 6, \ldots) \quad (4.1.27)$$

Comparing (4.1.21) and (4.1.23) with (3.5.8), we deduce that when attributing a wave function to a particle entails automatically, it is submitted to boundary conditions obtaining for k equivalent expressions with those of the guided electromagnetic waves.

4.2. Harmonic Oscillator in Quantum Mechanics

• From the classical expressions to the quantum mechanical ones

In order to have a precise understanding of the mathematical process of the quantization of the electromagnetic field, with which we will deal in the next chapters, it is of outstanding importance to know how the harmonic oscillator is described in the quantum mechanical approach.

We may start with the simple case of one dimensional motion along the x axis of a particle of mass m submitted to a central force expressed by the well-known Hooke's law

$$\vec{F} = -\frac{\partial V(x)}{\partial x} \frac{\vec{x}}{|x|} = -K_x \vec{x} \quad (4.2.1)$$

where $V(x)$ is the potential energy and K_x is the force constant.

$$V(x) = \frac{1}{2}K_x x^2 \qquad (4.2.2)$$

The kinetic energy at any coordinate x of the motion of the particle is

$$T_x = \frac{1}{2}m\left(\frac{\partial x}{\partial t}\right)^2 = \frac{P_x^2}{2m} \qquad (4.2.3)$$

where $P_x = m(\frac{\partial x}{\partial t})$ is the momentum of the particle along the x axis.
The total energy of the system writes

$$E^{(x)} = T_x + V(x) = \frac{P_x^2}{2m} + \frac{1}{2}K_x x^2 \qquad (4.2.4)$$

Consequently, putting

$$\omega_x = \left(\frac{K_x}{m}\right)^{1/2} \qquad (4.2.5)$$

the energy simply becomes

$$E^{(x)} = \frac{P_x^2}{2m} + \frac{1}{2}m\omega_x^2 x^2 \qquad (4.2.6)$$

Notice that ω_x has the dimensions of an angular frequency along the x axis.

As we have seen in the previous chapter, the general expression of the three dimensions momentum operator in quantum mechanics is

$$\tilde{P} = -i\,\hbar\vec{\nabla} \qquad (4.2.7)$$

which writes along the x axis

$$\tilde{P}_x = -i\,\hbar\frac{\partial}{\partial x} \qquad (4.2.8)$$

Considering $\psi(x)$ to be the eigenfunction, the Schrödinger equation for the linear harmonic oscillator derives immediately from (4.2.6),

$$-\frac{\hbar^2}{2m}\frac{\partial^2\psi(x)}{\partial x^2} + \frac{1}{2}m\omega_x^2 x^2\psi(x) = E^{(x)}\psi(x) \qquad (4.2.9)$$

Putting

$$\xi = x \left(\frac{mK_x}{\hbar^2} \right)^{1/4} \tag{4.2.10}$$

and considering the dimensionless fraction

$$\eta_x = \frac{2E^{(x)}}{\hbar\omega_x} \tag{4.2.11}$$

we get the simplified equation

$$\frac{\partial^2 \psi(\xi)}{\partial \xi^2} + (\eta_x - \xi^2)\psi(\xi) = 0 \tag{4.2.12}$$

The asymptotic behavior of the above equation when $|\xi| \to \infty$ suggests looking for solutions of the form:

$$\psi(\xi) = e^{-\xi^2/2} H(\xi) \tag{4.2.13}$$

where $H(\xi)$ should be polynomial functions not affecting the asymptotic behavior.

Solutions of (4.2.12) are found for

$$\eta_x = 2n_x + 1 \quad \text{with } \eta_x = 0, 1, 2, \ldots \tag{4.2.14}$$

called the Hermite polynomials $H_n(\xi)$ whose general expression is

$$H_{n_x}(\xi) = (-1)^{n_x} e^{\xi^2} \frac{d^{n_x} e^{-\xi^2}}{d\xi^{n_x}} \tag{4.2.15}$$

Getting the complete eigenfunctions

$$\psi_{n_x}(\xi) = C e^{-\xi^2/2} H_{n_x}(\xi) = C(-1)^{n_x} e^{\xi^2/2} \frac{d^{n_x} e^{-\xi^2}}{d\xi^{n_x}} \tag{4.2.16}$$

Where C is a constant which can be determined by imposing the summation over the whole space of the square modulus of the wave function to be normalized to unity. And the eigenvalues, that is the energy spectrum with $E^{(x)} \to E_{n_x}(\omega_x)$ writes

$$E_{n_x}(\omega_x) = \hbar\omega_x \left(n_x + \frac{1}{2} \right) \tag{4.2.17}$$

Hence, the energy of the linear one dimensional harmonic oscillator in quantum mechanics can only get the discrete values of relation (4.2.17) with $n_x = 0, 1, 2, \ldots$.

This is easily generalized in a three dimensional harmonic oscillator for which the Schrödinger equation and the energy spectrum take the expressions

$$-\frac{\hbar^2}{2m}\vec{\nabla}^2\Psi(\vec{r}) + \frac{1}{2}m\left[w_x^2 x^2 + w_y^2 y^2 + w_z^2 z^2\right]\Psi(\vec{r}) = E\,\Psi(\vec{r})$$

$$(4.2.18)$$

$$E = \hbar w_x\left(n_x + \frac{1}{2}\right) + \hbar w_y\left(n_y + \frac{1}{2}\right) + \hbar w_z\left(n_z + \frac{1}{2}\right) \quad (4.2.19)$$

This is a quite amazing result showing that a zero energy state cannot exist, and even an one dimensional harmonic oscillator, in complete absence of any external interaction, at zero temperature, has a minimum nonzero energy which equals to $E_0(w) = \frac{1}{2}\hbar w$ depending only on the potential and the mass of the particle according to (4.2.5).

- Dirac representation, creation and annihilation operators, a, a^+

For an one dimensional harmonic oscillator of mass m at a coordinate $\vec{q}$ with a momentum $\vec{p} = m\,d\vec{q}/dt$, employing the reduced expressions for the canonical variables of position $Q = |\vec{q}|\sqrt{m}$ and momentum $P = |\vec{p}|/\sqrt{m}$, the total energy (4.2.6) becomes

$$E = \frac{1}{2}(P^2 + w^2 Q^2) \quad (4.2.20)$$

The transition to the quantum mechanical Hamiltonian is obtained by imposing the following expressions for the momentum and position operators respectively

$$\hat{P} = i\sqrt{\frac{\hbar w}{2}}(a^+ - a) \quad \text{and} \quad \hat{Q} = \sqrt{\frac{\hbar}{2w}}(a^+ + a) \quad (4.2.21)$$

where a, a^+ are the annihilation and creation operators of a quantum of energy of the harmonic oscillator, which can be expressed by inversing equation (4.2.21)

$$a = \sqrt{\frac{1}{2\hbar w}}(w\hat{Q} + i\hat{P}) \quad \text{and} \quad a^+ = \sqrt{\frac{1}{2\hbar w}}(w\hat{Q} - i\hat{P}) \quad (4.2.22)$$

Considering (4.1.15), the commutation relation between the position and momentum operators is

$$[Q, P] = i\hbar \tag{4.2.23}$$

and we get

$$[a, a^+] = 1 \tag{4.2.24}$$

Considering (4.2.21) and (4.2.20), we obtain directly the Hamiltonian of the harmonic oscillator expressed with respect to the annihilation and creation operators

$$\tilde{H}_{HO} = \frac{1}{2}(\hat{P}^2 + \omega^2 \hat{Q}^2) = \hbar\omega \left(a^+ a + \frac{1}{2} \right) \tag{4.2.25}$$

In Dirac representation, each eigenfunction takes the simple expression $\psi_n(\xi) \Rightarrow |n\rangle$ corresponding to a quantum state composed of n harmonic oscillators.

It can be shown that the actions of the creation and annihilation operators a^+ and a respectively on the eigenfunction are expressed as following

$$a^+|n\rangle = \sqrt{n+1}|n+1\rangle$$
$$a|n\rangle = \sqrt{n}|n-1\rangle \tag{4.2.26}$$

Consequently, the simplified Schrödinger equation in Dirac representation is

$$\tilde{H}_{HO}|n\rangle = E|n\rangle \rightarrow \hbar\omega \left(a^+ a + \frac{1}{2} \right) |n\rangle$$

$$= \hbar\omega \left(n + \frac{1}{2} \right) |n\rangle \tag{4.2.27}$$

The successive application of a first and a^+ next, i.e., $a^+ a$ on the $|n\rangle$ state translates the number n of the quantum oscillators having energy $\hbar\omega$. Thus, $a^+ a$ is also called the number operator.

4.3. Quantum Electrodynamics (QED) and the Photon Description

Brief description of selected experiments that have historically played an important role for the introduction of the photon concept.

We have seen in 4.1 that for the interpretation of black body radiation, Planck was the first to introduce the revolutionary idea of the quantized absorptions and emissions of the electromagnetic field by oscillators.

In this chapter, we will discuss the main experiments that have led to the particle concept of the electromagnetic field, and then we will describe the theoretical methods that have been developed, in order to transit from the continuous electromagnetic waves to the quantum electromagnetic entity called the photon.

• The photoelectric effect and the quantum interpretation

By the end of the 19th century, it has been observed by many physicists like Hertz, Lenard, Stoletov and others, that metal surfaces eject charged particles when irradiated by ultraviolet light. In a series of experiments measuring the charge to mass ratio of the ejected charged particles, Lenard demonstrated that these were electrons and the effect was called photoelectric.

Based on Planck's ideas on the quantized emission of radiation by the equally quantized energy levels of atomic oscillators for the interpretation of the black body radiation, Einstein proposed in 1905, an explanation for the photoelectric effect. He advanced that light consists of point particles, called photons, carrying an amount of energy proportional to the frequency of the light $E = h\nu$, where h is Planck's constant.

Hence, considering that a minimum work W is required for an electron to escape from the surface of the metal, he deduced that the maximum kinetic energy of the ejected electron is simply obtained by

$$E_{\text{kin}} = \frac{mu^2}{2} = h\nu - W \tag{4.3.1}$$

From 1914 to 1917, Millikan carried out many experiments on the photoelectric effect using the visible light on lithium, potassium and sodium, all of these characterized by low values of the extraction work W and confirmed the last relation.

- Compton scattering

Modifications of the wavelength of X-rays when scattered by electrons have been reported by many experiments between 1912 and 1920. In 1922, for the interpretation of the experimental observations, Compton, who had carried out many similar experiments himself, proposed the revolutionary concept that X-rays whose wavelength is modified are scattered in crystals by electrons almost at rest.

The theoretical aspect of this effect is described by considering not only electrons but also photons as integral particles and by applying the energy and momentum conservation laws.

Consider an X-ray photon with frequency ν colliding with an electron with rest mass m_0 which after the collision acquires momentum.

The energy conservation law writes

$$h\nu + m_0 c^2 = h\nu' + mc^2 \Rightarrow h(\nu - \nu') = (m - m_0)c^2 \qquad (4.3.2)$$

and the momentum conservation

$$\frac{h\nu}{c} = \frac{h\nu'}{c}\cos\theta + p_e\cos\phi \Rightarrow \frac{h\nu}{c} - \frac{h\nu'}{c}\cos\theta = p_e\cos\phi \qquad (4.3.3)$$

$$\frac{h\nu'}{c}\sin\theta = p_e\sin\phi \Rightarrow \left(\frac{h\nu'}{c}\right)^2\sin^2\theta = p_e^2\sin^2\phi \qquad (4.3.4)$$

where θ is the angle of the scattered photon trajectory from the collision axis and ϕ that of the scattered electron.

Taking the square of (4.3.3) and using (4.3.4) and (4.3.2), we get

$$p_e^2 = \left(\frac{h\nu}{c}\right)^2 - 2\frac{h\nu}{c}\frac{h\nu'}{c} + \left(\frac{h\nu'}{c}\right)^2 + 2m_0 c\frac{h(\nu - \nu')}{c} \qquad (4.3.5)$$

The energy conservation for the electron writes

$$m^2 c^4 = m_0^2 c^4 + p_e^2 c^2 \Rightarrow p_e^2 = (m^2 - m_0^2)c^2 \qquad (4.3.6)$$

Combining (4.3.3) and (4.3.4), we obtain

$$p_e^2 = \left(\frac{h\nu}{c}\right)^2 - 2\frac{h\nu}{c}\frac{h\nu'}{c}\cos\theta + \left(\frac{h\nu'}{c}\right)^2 \qquad (4.3.7)$$

Equating (4.3.5) and (4.3.7), we get

$$\left(\frac{h\nu'}{c}\right)^2 + \left(\frac{h\nu}{c}\right)^2 - 2\frac{h\nu}{c}\frac{h\nu'}{c}\cos\theta$$

$$= \left(\frac{h\nu}{c}\right)^2 - 2\frac{h\nu}{c}\frac{h\nu'}{c} + \left(\frac{h\nu'}{c}\right)^2 + 2m_0c\frac{h(\nu - \nu')}{c}$$

getting the frequency of the scattered photon by the electron

$$\nu' = \frac{\nu}{1 + \frac{h\nu(1-\cos\theta)}{m_0c^2}} \tag{4.3.8}$$

The last relation obtained by particle dynamics interprets quite satisfactorily the experimental results demonstrating the particle nature of the photon, which is characterized by energy, momentum and directionality.

Finally, what is also extremely important here is that the Compton scattering demonstrates that the energy of a single photon, by the same token, its wavelength and frequency, can be modified during a collisional process.

• Low intensity Young's double-slit interferences

As we have seen in the historical survey, Young performed the famous double-slit experiment in 1802, obtaining an interference pattern that could be only explained by the wave theory of the light. At that time, it was the strongest argument supporting the wave concept of the light against Newton's particle hypothesis.

However, in 1909, Taylor repeated the same experiments by using extremely low intensity light sources (Fig. 4.3.1) and observed that "spots" were appearing on a photographic plate whose graduate accumulation tended to form interference patterns.

This was readily interpreted by many authors as a demonstration of the photons existence. More recently, using lasers and single photon pulse techniques, many experiments have also revealed the creation of spots on the detection screen and the gradual formation of interference pattern.

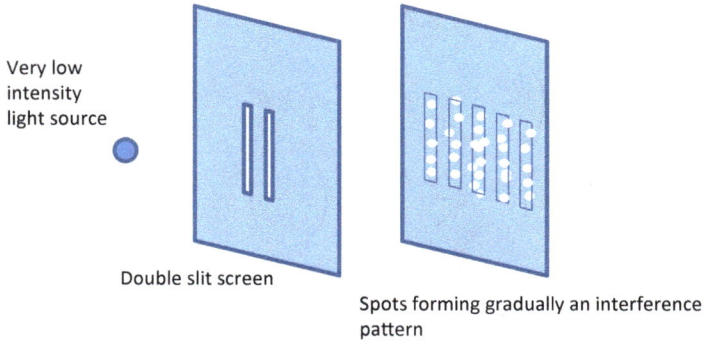

Fig. 4.3.1. Schematic representation of Taylor's double slit diffraction experiment in which the light source has an extremely low intensity.

4.3.1. *Second Quantization*

The photoelectric effect, Compton's scattering and Taylor's experiments, among others, have given a decisive impulse toward the general acceptance of the particle nature of light and consequently, a theoretical effort had started to model the quantized electromagnetic field based upon the harmonic oscillator quantum mechanical model described in (4.2).

- Quantization process of the electromagnetic field

   We will give now the main lines of the electromagnetic field quantization procedure, which is generally described in the Coulomb gauge.

   Considering that the electromagnetic field vector potential $\vec{A}(\vec{r}, t)$ consists of a superposition of plane waves, the classical electromagnetic field energy in a given volume V can be written according to (3.4.8) as

$$E_V = 2\varepsilon_0 V \sum_{k,\lambda} \omega_k^2 |A_{k\lambda}|^2 \qquad (4.3.9)$$

where ε_0 is the vacuum permittivity and $|A_{k\lambda}|$ the vector potential amplitude corresponding to the mode with a wave vector k, angular frequency ω_k and polarization λ. Caution has to be taken here not to confuse two different physical entities, number of polarizations and wavelength, which are both denoted in QED with λ.

Notice that (4.3.9) derives from (3.4.8), which is a mean value over a period, in other words over the wavelength. It is generally accepted that it is difficult to conceive a photon along the propagation axis with dimensions less than the wavelength. On the other hand, all the calculations on the guided propagation of the electromagnetic waves developed in Chapter 3 (3.5) show that a minimum volume has to be considered for a given mode to subsist. Consequently, caution has to be taken when using the formalism (4.3.9) in a volume V of given dimensions for the total energy calculation based on the modes summation.

In fact, for a fixed volume V we can write more rigorously

$$E_{EM} = 2\varepsilon_0 V \sum_{k \geq k_c, \lambda} \omega_k^2 |A_{k\lambda}|^2 \tag{4.3.10}$$

where the summation on the wave vectors has a lower cut-off limit defined by the boundary conditions of the box dimensions V. For the sake of simplicity we will drop this lower limit in the next calculations, which comes out to consider that the volume V is much bigger than the wavelengths of the modes considered.

The equivalence with the harmonic oscillator Hamiltonian is obtained by introducing equivalent expressions with the definitions (4.2.21) and (4.2.22), relating $A_{k\lambda}$ to the canonical variables of position $Q_{k\lambda}$ and momentum $P_{k\lambda}$

$$A_{k\lambda} = \frac{1}{2\omega_k \sqrt{\varepsilon_0 V}} (\omega_k Q_{k\lambda} + i P_{k\lambda})$$

$$A_{k\lambda}^* = \frac{1}{2\omega_k \sqrt{\varepsilon_0 V}} (\omega_k Q_{k\lambda} - i P_{k\lambda})$$

$$Q_{k\lambda} = \sqrt{\varepsilon_0 V} (A_{k\lambda} + A_{k\lambda}^*)$$

$$P_{k\lambda} = -i\omega_k \sqrt{\varepsilon_0 V} (A_{k\lambda} - A_{k\lambda}^*) \tag{4.3.11}$$

Using the above relations in (4.3.9), the energy of the electromagnetic field writes as a function of $Q_{k\lambda}$ and $P_{k\lambda}$

$$E_{EM} = \frac{1}{2} \sum_{k,\lambda} (P_{k\lambda}^2 + \omega_k^2 Q_{k\lambda}^2) \tag{4.3.12}$$

Comparing the last relation with (4.2.25), it can be deduced directly that each mode of the radiation field can be considered equivalent to a harmonic oscillator.

In order to express the amplitude of the vector potential with respect to the quantized energy of the photons $\hbar\omega$, we start from the monochromatic electromagnetic wave energy density equation (3.4.7). The energy density of N photons in a box V equals the mean energy density over a period $2\pi/\omega$ of the classical electromagnetic field

$$\langle 4\varepsilon_0\omega^2 A_0^2(\omega)\sin^2(\vec{k}\cdot\vec{r}-\omega t)\rangle_{\text{period}} = 2\varepsilon_0\omega^2 A_0^2(\omega) = \frac{N\hbar\omega}{V}$$

(4.3.13)

Consequently, for $N = 1$ in the relation (4.3.13), we obtain the link between the classical electromagnetic wave issued from Maxwell's equations and the quantized field operators used in QED, which is expressed by the intermediate of the vector potential amplitude in a volume V as follows

$$A_{k\lambda} \rightarrow \sqrt{\frac{\hbar}{2\varepsilon_0\omega_k V}}a_{k\lambda} \quad A_{k\lambda}^* \rightarrow \sqrt{\frac{\hbar}{2\varepsilon_0\omega_k V}}a_{k\lambda}^+$$

(4.3.14)

Where $a_{k\lambda}$ and $a_{k\lambda}^+$ are respectively the annihilation and creation operators of a k mode and λ polarization photon composing the photon number operator $\hat{N}_{k\lambda} = a_{k\lambda}^+ a_{k\lambda}$.

Using the above expressions for the amplitude of the vector potential the classical formalism of equation (3.4.2) becomes in QED, when taking into account all k modes and polarizations λ,

$$\vec{A}(\vec{r},t) = \sum_{k,\lambda}\sqrt{\frac{\hbar}{2\varepsilon_0\omega_k V}}[a_{k\lambda}\hat{\varepsilon}_{k\lambda}e^{i(\vec{k}\cdot\vec{r}-\omega_k t+\theta)} + a_{k\lambda}^+\hat{\varepsilon}_{k\lambda}^*e^{-i(\vec{k}\cdot\vec{r}-\omega_k t+\theta)}]$$

(4.3.15)

According to (4.2.26), the creation and annihilation operators of a k mode and λ polarization photon have the properties

$$a_{k\lambda}^+|n_{k\lambda}\rangle = \sqrt{n_{k\lambda}+1}|n_{k\lambda}+1\rangle$$
$$a_{k\lambda}|n_{k\lambda}\rangle = \sqrt{n_{k\lambda}}|n_{k\lambda}-1\rangle$$
$$a_{k\lambda}^+ a_{k\lambda}|n_{k\lambda}\rangle = \hat{N}_{k\lambda}|n_{k\lambda}\rangle = n_{k\lambda}|n_{k\lambda}\rangle$$

(4.3.16)

where $\hat{N}_{k\lambda}$, the photon number operator expresses the number of photons.

Using the link expressions (4.3.14) in the equations of $Q_{k\lambda}$ and $P_{k\lambda}$ given in (4.3.11), one obtains the position and momentum operators expressed with respect to $a_{k\lambda}$ and $a_{k\lambda}^+$

$$\hat{Q}_{k\lambda} = \sqrt{\frac{\hbar}{2\omega_k}}(a_{k\lambda}^+ + a_{k\lambda}) \quad \hat{P}_{k\lambda} = i\sqrt{\frac{\hbar\omega_k}{2}}(a_{k\lambda}^+ - a_{k\lambda}) \quad (4.3.17)$$

Replacing the canonical variables $Q_{k\lambda}$ and $P_{k\lambda}$ in equation (4.3.12) by the expressions of $\hat{Q}_{k\lambda}$ and $\hat{P}_{k\lambda}$ (4.3.17) and considering the commutation relation $[a_{k\lambda}, a_{k\lambda}^+] = 1$ we get the Hamiltonian of the quantized electromagnetic field

$$H_{EM} = \frac{1}{2}\sum_{k,\lambda}(\hat{P}_{k\lambda}^2 + \omega_k^2\hat{Q}_{k\lambda}^2) = \sum_{k,\lambda}\hbar\omega_k\left(\hat{N}_{k\lambda} + \frac{1}{2}\right) \quad (4.3.18)$$

Hence, in QED, photons are considered as point particles created and annihilated by the operators $a_{k\lambda}^+$ and $a_{k\lambda}$ respectively which obey the properties of (4.3.16), and having the energy of an ensemble of harmonic oscillators given by (4.3.18).

Generally, the polarization λ takes two values corresponding to Right- and Left-hand circularly polarized photons.

The term $1/2$ of the established radiation Hamiltonian is an occupation operator, which rigorously means that in complete absence of photons, i.e., when $\hat{N}_{k\lambda} = 0$, empty space is filled permanently by "half" photons of all frequencies and polarizations. Hence, the quantization procedure of the classical expression of the energy of the electromagnetic waves to get the radiation Hamiltonian leads to the zero-point radiation field, also called the quantum vacuum, expressed by the term $\sum_{k,\lambda}\frac{1}{2}\hbar\omega_k$. This term is mainly responsible for the QED singularities related to the infinite energy of the vacuum, while at the same time it is pretexted that it provides the basis for the explanation of various experimentally observed effects, such as spontaneous emission, atomic levels energy shifts (Lamb shift), the Casimir effect etc.

4.4. Interaction between Electromagnetic Waves and Charged Particles, Reality of the Vector Potential

• Interaction Hamiltonian between an electromagnetic wave and a charged particle

The interaction Hamiltonian of a charged particle of charge q and mass m, having a momentum $\vec{p}$, in an electromagnetic field characterized by a vector potential $\vec{A}(\vec{r},t)$ and a scalar potential $\Phi(\vec{r},t)$ is written as

$$H = \frac{1}{2m}(\vec{p} - q\vec{A}(\vec{r},t))^2 + q\Phi(\vec{r},t) \qquad (4.4.1)$$

Using the Coulomb gauge

$$\vec{\nabla} \cdot \vec{A}(\vec{r},t) = 0 \quad \text{and} \quad \Phi(\vec{r},t) = 0 \qquad (4.4.2)$$

and considering the momentum operator (4.2.7), the time dependent Schrödinger equation for an electron, where $q = -e$, in the electrostatic Coulomb potential of an hydrogenic atom $-\frac{Ze^2}{4\pi\varepsilon_0 r}$, interacting with the electromagnetic field writes

$$i\hbar\frac{\partial}{\partial t}\Psi(\vec{r},t)$$
$$= \left(-\frac{\hbar^2}{2m}\vec{\nabla}^2 - \frac{Ze^2}{4\pi\varepsilon_0 r} - \frac{i\hbar e}{m}\vec{A}(\vec{r},t)\cdot\vec{\nabla} + \frac{e^2}{2m}\vec{A}(\vec{r},t)^2\right)\Psi(\vec{r},t)$$
$$(4.4.3)$$

In fact, it can be easily shown that the Coulomb gauges makes $\vec{A}$ and $\vec{\nabla}\cdot$ to commute

$$\vec{\nabla}\cdot(\vec{A}\Psi) = \vec{A}\cdot(\vec{\nabla}\Psi) + (\vec{\nabla}\cdot\vec{A})\Psi = \vec{A}\cdot(\vec{\nabla}\Psi) \qquad (4.4.4)$$

In the case of weak fields, the term of the square of the vector potential in Schrodinger equation is quite small compared to the linear one and consequently, the main part of the interaction Hamiltonian between the bounded electron and the electromagnetic field writes

$$H_{\text{int}}(t) \cong -\frac{i\hbar e}{m}\vec{A}(\vec{r},t)\cdot\vec{\nabla} \qquad (4.4.5)$$

which is time dependent because of the vector potential.

From the established interaction Hamiltonian, arises the question whether the vector potential represents a mathematical artifact or if it corresponds to a physical entity.

- Reality of the vector potential, Ehrenberg–Siday or Aharonov–Bohm effect.

Obviously, according to the Hamiltonian (4.4.1), the vector potential directly affects the modulus and the direction of the momentum of a charged particle.

In 1949, W. Ehrenberg and R.E. Siday were the first to deduce the effect of the vector potential of the electromagnetic fields on charged particles. The Ehrenberg–Siday effect is a fine demonstration of the real existence of the scalar and vector fields issued from the electromagnetic theory. In fact, before Ehrenberg's and Siday's works, the vector and scalar fields, $\vec{A}(\vec{r}, t)$; $\Phi(\vec{r}, t)$ respectively, were believed to be simply mathematical constructs for the development of the gauge theories in electromagnetism. In reality, both of these potentials interact with a charged particle even in absence of electric and magnetic fields.

The same effects were later rediscovered by Aharonov and Bohm in 1959 and were confirmed experimentally by Chambers in 1960 and Osakabe in 1986 using a quite advanced experimental apparatus.

4.5. Transition Rates and Vacuum Induced Spontaneous Emission

For describing the transitions between the quantized energy levels of the atoms, we have started in the previous chapter, defining the interaction Hamiltonian (4.4.1) of a charged particle in an electromagnetic field characterized by a vector potential $\vec{A}(\vec{r}, t)$ and a scalar potential $\Phi(\vec{r}, t)$.

Hence, the interaction Hamiltonian is a time dependent perturbation and can be studied within the time dependent perturbation theory according to which the wave function $\Psi(\vec{r}, t)$ is expended in terms of the unperturbed eigenfunctions $\varphi_l(\vec{r})$, solutions of the time independent Schrödinger equation

$$H_0 \varphi_l(\vec{r}, t) = \left(-\frac{\hbar^2}{2m} \vec{\nabla}^2 - \frac{Ze^2}{4\pi\varepsilon_0 r} \right) \varphi_l(\vec{r}, t) = E_l \varphi_l(\vec{r}, t) \qquad (4.5.1)$$

The expansion of the wave function writes

$$\Psi(\vec{r},t) = \sum_l c_l(t)\varphi_l(\vec{r})e^{-iE_l t/\hbar} \qquad (4.5.2)$$

where the summation runs over all the discrete and continuous sets of the eigenfunctions $\varphi_l(\vec{r})$. The time dependence is included in the exponential argument and in the coefficients $c_l(t)$, which in the time dependent perturbation theory satisfy the coupled equations

$$i\hbar\frac{dc_m^{(n+1)}(t)}{dt} = \sum_l \int \varphi_m^*(\vec{r})H_{\text{int}}(t)\varphi_i(\vec{r})d^3r\, c_l^{(n)}(t)e^{i\omega_{ml}t} \qquad (4.5.3)$$

where $n = 0, 1, 2, \ldots$ and $E_m - E_l = \hbar\omega_{ml}$

Assuming that the system is initially at $t = 0$ in a defined stationary state $\varphi_i(\vec{r})$ corresponding to the energy E_i, for $n = 1$, the coefficient $c_l^{(0)}(t) = \delta_{li}$ for the discrete states or $c_l^{(0)}(t) = \delta(l-i)$ for the continuous ones.

Where, δ_{li} is Kronecker delta symbol, $\delta_{li} = 1$ if $l = i$ and $\delta_{li} = 0$ if $l \neq i$ and $\delta(l-i)$ is Dirac delta function, $\delta(0) = 1$ (if $l = i$) and $\delta(l-i) = 0$ if $l \neq i$

In this case the last equation gives

$$i\hbar\frac{dc_m^{(1)}(t)}{dt} = \int \varphi_m^*(r)H_{\text{int}}(t)\varphi_i(r)\, d^3r e^{i\omega_{mi}t} \qquad (4.5.4)$$

Using the interaction Hamiltonian (4.4.5) and the expression for the vector potential (3.4.2), we obtain the transition probability amplitude at a given instant t by integrating within an angular frequency interval $\delta\omega$ and over the time interval from 0 to t

$$c_m^{(1)}(t) = -\frac{e}{m}\int_{\delta\omega} A_0(\omega)d\omega$$

$$\times \left[e^{i\theta}\langle\varphi_m|e^{i\vec{k}\cdot\vec{r}}\hat{\varepsilon}\cdot\vec{\nabla}|\varphi_i\rangle \int_0^t e^{i(\omega_{mi}-\omega)t'}dt' \right.$$

$$\left. + e^{-i\theta}\langle\varphi_m|e^{-i\vec{k}\cdot\vec{r}}\hat{\varepsilon}^*\cdot\vec{\nabla}|\varphi_i\rangle \int_0^t e^{i(\omega_{mi}+\omega)t'}dt' \right] \qquad (4.5.5)$$

The square modulus of $c_m^{(1)}(t)$ represents the probability of the system to be in the stationary state m at time t.

$$|c_m^{(1)}(t)|^2 = 2 \left(\frac{e}{m}\right)^2 \int_{\delta\omega} |A_0(\omega)|^2 \Theta(t, \omega - \omega_{\mathrm{mi}})|M_{\mathrm{mi}}(\omega)|^2 d\omega$$

(4.5.6)

where the matrix element $|M_{\mathrm{mi}}(\omega)|$ has the form

$$|M_{\mathrm{mi}}(\omega)| = \left| \int \varphi_m^*(\vec{r}) e^{i\vec{k}\cdot\vec{r}} \hat{\varepsilon} \cdot \vec{\nabla}\varphi_i(\vec{r}) d^3r \right|$$

(4.5.7)

The trigonometric function $\Theta(t, \omega - \omega_{\mathrm{mi}})$ involved in the integration of (4.5.6) results from the time integration of the exponentials and has the form

$$\Theta(t, \omega - \omega_{\mathrm{mi}}) = \frac{1 - \cos\left[(\omega - \omega_{\mathrm{mi}})t\right]}{(\omega - \omega_{\mathrm{mi}})^2}$$

(4.5.8)

Since $\Theta(t, \omega - \omega_{mi})$ takes its maximum value for $\omega = \omega_{mi}$ and its integration over all the frequencies gives πt we can consider in a first approximation that for a large time interval $\omega = \omega_{mi}$ in the expressions of the vector potential amplitude $A_0(\omega_{mi})$ and the matrix element between the states $\varphi_m(\vec{r})$ and $\varphi_i(\vec{r})$, $M_{\mathrm{mi}}(\omega_{\mathrm{mi}})$ and the integration of (4.5.6) now gives

$$|c_m^{(1)}(t)|^2 = 2\pi \left(\frac{e}{m}\right)^2 |A_0(\omega_{\mathrm{mi}})|^2 |M_{\mathrm{mi}}(\omega_{\mathrm{mi}})|^2 t$$

(4.5.9)

which increases linearly with time.

We obtain immediately the transition rates T for absorption

$$T_{\mathrm{abs}} = \frac{d}{dt}|c_m^{(1)}(t)|^2 = 2\pi \left(\frac{e}{m}\right)^2 |A_0(\omega_{\mathrm{mi}})|^2 |M_{\mathrm{mi}}(\omega_{\mathrm{mi}})|^2$$

(4.5.10)

and stimulated emission

$$T_{\mathrm{emi}}^{\mathrm{stimulated}} = \frac{d}{dt}|c_m^{(1)}(t)|^2 = 2\pi \left(\frac{e}{m}\right)^2 |A_0(\omega_{\mathrm{mi}})|^2 |M_{\mathrm{im}}(\omega_{\mathrm{mi}})|^2$$

(4.5.11)

for which the main difference relies on the order inversion of the matrix element between the initial and final state.

It is worth noting that for the above calculations, we have used the classical expression of the vector potential given by (3.4.2).

• Photoelectric effect and the semi-classical interpretation

At that level, within a chapter that deals mainly with the photon, we will make a parenthesis to present the way Wentzel and Beck in 1926, Lamb and Scully in 1960s demonstrated that the photoelectric effect may quite well be interpreted using the electromagnetic wave nature of the light.

In fact, we can calculate the transition rate in the case of the photoelectric effect from (4.5.6) and (4.5.8), exactly in the same way as in the previous paragraph

$$|c_m^{(1)}(t)|^2 = 2\left(\frac{e}{m}\right)^2 \int |A_0(\omega)|^2 \frac{1 - \cos[(\omega - \omega_{\mathrm{mi}})t]}{(\omega - \omega_{\mathrm{mi}})^2}|M_{\mathrm{mi}}(\omega)|^2 d\omega$$

$$(4.5.12)$$

On the other hand we have

$$\int_{-\omega}^{+\omega} \frac{1 - \cos[(\omega - \omega_{\mathrm{mi}})t]}{(\omega - \omega_{\mathrm{mi}})^2} d\omega = \pi t \qquad (4.5.13)$$

And for sufficiently long time-periods

$$\lim_{t \to \infty} \frac{1 - \cos[(\omega - \omega_{\mathrm{mi}})t]}{(\omega - \omega_{\mathrm{mi}})^2} = \pi t \delta(\omega - \omega_{\mathrm{mi}}) \qquad (4.5.14)$$

Consequently (4.5.12) becomes

$$|c_m^{(1)}(t)|^2 = 2\pi t \left(\frac{e}{m}\right)^2 |A_0(\omega)|^2 |M_{\mathrm{mi}}(\omega)|^2 \delta(\omega - \omega_{\mathrm{mi}}) \qquad (4.5.15)$$

And the transition rate writes

$$T_{\mathrm{photoelectric}} = \frac{d}{dt}|c_m^{(1)}(t)|^2 = 2\pi \left(\frac{e}{m}\right)^2 |A_0(\omega)|^2 |M_{\mathrm{mi}}(\omega)|^2 \delta(\omega - \omega_{\mathrm{mi}})$$

$$(4.5.16)$$

Between an initial bounded state and the continuum level, Dirac delta function guarantees that the energy of the photoelectron satisfies the photoelectric equation. The rate of the emission is directly proportional to the square modulus of the radiation vector potential, and consequently to the field intensity. Obviously, because of (4.5.14), the calculated transition rate for the photoelectric effect is valid for time periods $t \gg \omega^{-1}$ and consequently for visible and UV light, it predicts the emission of electrons without any delay.

As remarked by many authors, a semi-classical calculation, in which the electromagnetic field is strictly considered with the classical wave expressions, gives an excellent interpretation of the photoelectric effect without the introduction of the photon concept. Recalling that the photoelectric effect is used experimentally to detect photons, this is a quite upsetting result.

• Spontaneous emission rate

As we have discussed in (4.3.18) in the QED description, the vacuum state is composed by an infinite quantity of photons of all frequencies and polarizations $\sum_{k,\lambda} \frac{1}{2}\hbar\omega_k$. Consequently, in order to describe the spontaneous emission we have to use the expression (4.3.15) for the vector potential, since in the classical electromagnetic theory the vacuum has no electromagnetic fields and consequently (3.4.2) is zero.

The part of the quantum expression of the vector potential responsible for the creation of a photon is

$$\langle \vec{A}_C(\vec{r}, t) \rangle = \langle \dots n_{k\lambda}, n_{k\lambda} + 1 \dots |$$

$$\times \sqrt{\frac{\hbar}{2\varepsilon_0 \omega_k V}} \left[a_{k\lambda}^+ \hat{\varepsilon}_{k\lambda}^* e^{-i(\vec{k}\cdot\vec{r} - \omega_k t + \theta)} \right] | \dots n_{k\lambda} \dots \rangle$$

$$= \sqrt{\frac{\hbar(n_{k\lambda} + 1)}{2\varepsilon_0 \omega_k V}} \left[\hat{\varepsilon}_{k\lambda}^* e^{-i(\vec{k}\cdot\vec{r} - \omega_k t + \theta)} \right] \qquad (4.5.17)$$

Of course, we can use the same expression for the transition rate as in (4.5.16), which writes here

$$T_{\text{emi}} = \pi \left(\frac{e}{m}\right)^2 \frac{\hbar(n_{k\lambda} + 1)}{\varepsilon_0 \omega_k V} |M_{\text{im}}|^2 \delta(\omega_k - \omega_{\text{mi}}) \qquad (4.5.18)$$

In absence of k mode and λ polarization photon, i.e., when $n_{k\lambda} = 0$, the transition rate does not vanish and develop within a direction

$$T_{\text{emi}}^{\text{spontaneous}} = \left(\frac{e}{m}\right)^2 \frac{\pi \hbar}{\varepsilon_0 \omega_k V} |M_{\text{im}}|^2 \delta(\omega_k - \omega_{\text{mi}}) \qquad (4.5.19)$$

which depends on the volume V of the system, but if we consider the density of states (3.5.42) and integrate over ω and all spatial directions of emission, while summing on both possible photon polarization ($\lambda = 1, 2$), then the parameter V is eliminated and we get a final expression of the total transition rate which is independent of V.

$$T_{\text{emi}}^{\text{spontaneous}}(total) = \left(\frac{e}{m}\right)^2 \frac{\hbar}{8\pi^2 \varepsilon_0 c^3} \int d\Omega \sum_{\lambda} |M_{\text{im}}|_{\lambda}^2 \omega_{\text{mi}} \quad (4.5.20)$$

The special interest on the spontaneous emission arises from the fact that it results from the coupling of the atomic levels to the quantum vacuum state.

• Dipole approximation and spontaneous emission

Within the dipole approximation, according to which when the wavelength of the electromagnetic field is too big with respect to the atomic dimension, i.e., when $kr \ll 1$, the exponential in (4.5.7) can be considered to be close to unity and using (4.2.7) the matrix element writes

$$M_{\text{mi}} = \int \varphi_m^*(\vec{r}) \hat{\varepsilon} \cdot \vec{\nabla} \varphi_i(\vec{r}) d^3 r = \hat{\varepsilon} \cdot \left(\frac{i}{\hbar}\right) \int \varphi_m^*(\vec{r}) \vec{p} \varphi_i(\vec{r}) d^3 r$$

$$(4.5.21)$$

According to Heisenberg equation of motion for a vector observable $\vec{F}$

$$\frac{d}{dt}\vec{F} = -\frac{i}{\hbar}[\vec{F}, H_0] \quad (4.5.22a)$$

$$\vec{p} = m\frac{d}{dt}\vec{r} = -m\frac{i}{\hbar}[\vec{r}, H_0] \quad (4.5.22b)$$

where H_0 is the unperturbed Hamiltonian of the system.

So (4.5.21) becomes

$$M_{\text{mi}} = \hat{\varepsilon} \cdot \left(\frac{m}{\hbar^2}\right) \int \varphi_m^*(\vec{r})[\vec{r}, H_0]\varphi_i(\vec{r}) d^3 r$$

$$= \hat{\varepsilon} \cdot \left(\frac{m}{\hbar^2}\right) (E_i - E_m) \int \varphi_m^*(\vec{r}) \vec{r} \varphi_i(\vec{r}) d^3 r$$

$$= -\left(\frac{m}{\hbar}\right) \omega_{\text{mi}} \hat{\varepsilon} \cdot \int \varphi_m^*(\vec{r}) \vec{r} \varphi_i(\vec{r}) d^3 r$$

$$= -\left(\frac{m}{\hbar}\right) \omega_{\text{mi}} \hat{\varepsilon} \cdot \vec{r}_{\text{mi}} = -\left(\frac{m}{\hbar}\right) \omega_{\text{mi}} |r_{\text{mi}}| \cos\theta \quad (4.5.23)$$

where θ is the angle between $\vec{r}$ and the polarization vector $\hat{\varepsilon}$.

Thus, the spontaneous emission rate writes

$$T_{\text{emi}}^{\text{spontaneous}}(\text{total}) = \frac{e^2}{8\pi^2 \hbar \varepsilon_0 c^3} \omega_{\text{mi}}^3 \int d\Omega \sum_\lambda |\vec{r}_{\text{mi}}|_\lambda^2 \cos^2\theta$$

$$= \frac{e^2}{3\pi \varepsilon_0 \hbar c^3} \omega_{\text{mi}}^3 |\vec{r}_{\text{mi}}|^2 \qquad (4.5.24)$$

where we have considered two polarizations and $\cos^2\theta$ has been replaced by the mean value

$$\langle\cos^2\theta\rangle = \frac{1}{4\pi} \int \cos^2\theta \, d\Omega = \frac{1}{4\pi} \int_0^{2\pi} d\phi \int_{-1}^{+1} \cos^2\theta \, d(\cos\theta) = \frac{1}{3}$$

$$(4.5.25)$$

Finally, it can be demonstrated that the interaction Hamiltonian between an electron and the electromagnetic field in the dipole approximation can be simply written

$$H_{\text{int}} = -\vec{D} \cdot \vec{E}(\vec{r}, t) = -e\vec{r} \cdot \vec{E}(\vec{r}, t) \qquad (4.5.26)$$

with $\vec{D} = e\vec{r}$ the electric dipole moment of the electron with charge e at the position $\vec{r}$ from the nucleus and $\vec{E}(\vec{r}, t)$, the electric field of the electromagnetic wave obtained by the vector potential according to (3.3.15).

- The photon spin

Conservation of angular momentum during spectroscopy experiments has shown that each photon has a component of angular momentum along the propagation axis of magnitude $\pm\hbar$. This corresponds to the intrinsic spin of photons whose component along the propagation axis is also called helicity. A left-hand circularly polarized photon has an helicity of $+\hbar$ while a right-hand circularly polarized photon has an helicity of $-\hbar$.

4.6. Lamb Shift

- Nonrelativistic calculations: Bethe's approach

The energy difference between the energy levels $2S_{1/2}$ and $2P_{3/2}$ of the atomic hydrogen was measured precisely by W.E. Lamb and

R.C. Retherford in 1947 using microwave technics. They obtained a value of 4.5×10^{-5} eV, which was not explained by Dirac's theory, which predict these two levels to have the same energy.

With the development of QED and the introduction of the notion of the quantum vacuum, many scientists advanced the hypothesis that the experimentally observed shifts of the atomic levels, named Lamb shifts, are due to the interaction of the bounded electrons with the vacuum. The first to propose a theoretical interpretation was H.A. Bethe in 1947 by introducing an astonishing way for extracting finite quantities from singularities. We reproduce briefly here those calculations.

Since the basis of the hypothesis is the interaction of an electron with the vacuum photons, we go back to Schrödinger equation (4.4.3) and start by examining the contributions of the vector potential terms.

The quadratic term of the vector potential contributes in the same manner to all atomic states since it does not contain atomic operators

$$\Delta E(\vec{A^2}) = \langle n_{k\lambda}, vacuum | \frac{e^2}{2m} \vec{A^2} | n_{k\lambda}, vacuum \rangle$$

$$= \frac{e^2}{2m} \sum_{k,\lambda} \frac{\hbar}{2\varepsilon_0 \omega_k V} = \frac{e^2 \hbar}{4\pi^2 \varepsilon_0 mc^3} \int_0^\infty \omega d\omega \quad (4.6.1)$$

where we have replaced the discrete summation by a continuous one according to the well- known transformation introduced in (3.5.44) and taking into account the polarization λ

$$\sum_{k,\lambda} \rightarrow \sum_{\lambda} \frac{V}{8\pi^3} \int d^3k = \frac{V}{8\pi^3} \sum_{\lambda} \int 4\pi k^2 dk = \frac{V}{2\pi^2 c^3} \sum_{\lambda} \int \omega^2 d\omega$$

$$(4.6.2)$$

The (4.6.1) expression is infinite. However, it is fully neglected based on the argument that it does not induce "measurable energy shifts".

Now, starting again from (4.4.3) and considering (4.5.15), the second order perturbation theory gives the energy shifts to the atomic levels, due to the coupling with the quantum vacuum resulting from the linear term of the vector potential in the interaction Hamiltonian

$$\Delta E_l(\vec{A}) = \sum_f \sum_{k,\lambda}$$

$$\times \frac{e^2}{m^2} \left(\frac{\hbar}{2\varepsilon_0 \omega_k V} \right) \frac{|\langle f, n_{k,\lambda} = 1 | a_{k,\lambda}^+ (\hat{\varepsilon}_{k,\lambda} \cdot \vec{p}) | l, n_{k\lambda} = 0 \rangle|^2}{E_l - E_f - \hbar\omega_k}$$

$$= \sum_f \sum_{k,\lambda} \frac{e^2}{m^2} \left(\frac{\hbar}{2\varepsilon_0 \omega_k V} \right) \frac{|\hat{\varepsilon}_{k,\lambda} \cdot \vec{p}_{fl}|^2}{E_l - E_f - \hbar\omega_k}$$

$$= \sum_f \frac{e^2}{4c^3 \varepsilon_0 m^2 \pi^2} \sum_\lambda |\vec{p}_{fl}|^2 \langle \cos^2 \theta \rangle \int \frac{\omega \, d\omega}{\left(\frac{E_l - E_f}{\hbar} \right) - \omega}$$

$$= \sum_f \left(\frac{e^2}{6\hbar m^2 c^3 \varepsilon_0 \pi^2} \right) |\vec{p}_{fl}|^2 \int \frac{E \, dE}{E_l - E_f - E}$$

$$= \sum_f \left(\frac{2\alpha_{\text{FS}}}{3\pi m^2 c^2} \right) |\vec{p}_{fl}|^2 \int_0^\infty \frac{E \, dE}{E_l - E_f - E} \qquad (4.6.3)$$

where we have used the relations (4.5.25) and (4.6.2) and considered two polarizations (right- and left-hand circular). Also α_{FS} is the fine structure constant

$$\alpha_{FS} = \frac{e^2}{4\pi\varepsilon_0 \hbar c} = \frac{1}{137.036}; \quad e^2 = 2\alpha_{FS} (\varepsilon_0 hc) \qquad (4.6.4)$$

As in the case of the quadratic term of the vector potential, the obtained expression (4.6.3) for the energy shift $\Delta E_l(\vec{A})$ induced by the linear part of the vector potential is also infinite.

To circumvent this difficulty, Bethe introduced a renormalisation procedure. In fact, for a free electron in the vacuum field, all the values of the atomic levels energy differences $E_l - E_f$ are zero so that $\Delta E_l(\vec{A})$ writes

$$\Delta E_l(\vec{A})_{\text{free}} = \sum_f \left(\frac{2\alpha_{\text{FS}}}{3\pi m^2 c^2} \right) |\vec{p}_{fl}|^2 \int_0^\infty dE \qquad (4.6.5)$$

in which the momentum operator between the atomic levels is preserved.

$\Delta E_l(\vec{A})_{\text{free}}$ is equally infinite. However, the experimentally observed shift for an atomic level $|l\rangle$ should be the difference

$$\Delta E_l(\vec{A}) - \Delta E_l(\vec{A})_{\text{free}}$$
$$= \sum_f \left(\frac{2\alpha_{FS}}{3\pi m^2 c^2} \right) |\vec{p}_{\text{fl}}|^2 (E_l - E_f) \int_0^\infty \frac{dE}{E_l - E_f - E} \quad (4.6.6)$$

This operation is named Bethe's renormalization and results to the relation (4.6.6) which is still infinite but through a slow logarithmic divergence.

Next, an upper level for the integration has been introduced and fixed to mc^2, m being the electron mass at rest, getting

$$\Delta E_l(\vec{A}) - \Delta E_l(\vec{A})_{\text{free}}$$
$$= \sum_f \left(\frac{2\alpha_{\text{FS}}}{3\pi m^2 c^2} \right) |\vec{p}_{\text{fl}}|^2 (E_f - E_l) \log \left| \frac{mc^2}{E_l - E_f} \right| \quad (4.6.7)$$

In order to get the logarithm out from the discrete summation, the mean value can be considered

$$\Delta E_l(\vec{A}) - \Delta E_l(\vec{A})_{\text{free}}$$
$$= \left(\frac{2\alpha_{\text{FS}}}{3\pi m^2 c^2} \right) \left\langle \log \left| \frac{mc^2}{E_l - E_f} \right| \right\rangle \sum_f |\vec{p}_{\text{fl}}|^2 (E_f - E_l) \quad (4.6.8)$$

where the average is calculated over all the discrete and continuous atomic levels

$$\left\langle \log \left| \frac{mc^2}{E_l - E_f} \right| \right\rangle \approx \log(mc^2) - \frac{\sum_f |\vec{p}_{\text{fl}}|^2 (E_f - E_l) \log |E_f - E_l|}{\sum_f |\vec{p}_{\text{fl}}|^2 (E_f - E_l)}$$
$$(4.6.9)$$

It can be easily shown that for a hydrogen atom with atomic number Z, we get

$$\sum_f |\vec{p}_{\text{fl}}|^2 (E_f - E_l) = 2\pi Z |e\hbar\varphi_l(o)|^2 \quad (4.6.10)$$

entailing that the shift should be more important for the s states for which $|\varphi_l(o)|^2 \neq 0$.

Thus (4.6.8) becomes

$$\Delta E_l(\vec{A}) - \Delta E_l(\vec{A})_{\text{free}} \approx \left(\frac{4\alpha_{\text{FS}} Z e^2 \hbar^2}{3 m^2 c^2} \right) \left\langle \log \left| \frac{mc^2}{E_l - E_f} \right| \right\rangle |\varphi_l(o)|^2$$

$$(4.6.11)$$

which is the calculated energy shift for the level $|l\rangle$.

Using a numerical estimate of the (4.6.9) along all the energy levels $|f\rangle$ up to the continuum spectrum, Bethe found for the 2S hydrogen level, an energy shift corresponding to a frequency of about 1000 MHz, in good agreement with the experimental value.

4.7. Conclusion Remarks

Some quite important concepts and calculation methodologies have been developed in this chapter.

The second quantization processes introduced the link equations (4.3.14) for the vector potential amplitude between the electromagnetic wave expressions and those of the QED creation and annihilation operators. This permitted to establish the correspondence between the electromagnetic wave and the harmonic oscillators in quantum mechanics establishing the Hamiltonian of the quantized electromagnetic field (4.3.18). The validity of this process will be discussed in the next chapters. Note that the link equations (4.3.14) etc., obtained by considering the mean value of the energy density of the electromagnetic wave over a period, in other words over a wavelength, are put equal to the value of a point photon with energy $\hbar\omega$ in a given volume V.

Furthermore, we have seen that the interaction Hamiltonian between the electromagnetic field and the charged particles is expressed through the vector potential, a real physical entity, first demonstrated by Ehrenberg and Siday. The same interaction Hamiltonian has been employed by Wentzel, Beck, Lamb and Scully to demonstrate that the photoelectric effect, generally advanced as a demonstration of the particle nature of the light, can quite well be interpreted through the electromagnetic wave representation.

Finally, we have briefly presented two important effects of the vacuum-electron interactions, the spontaneous emission and the

Lamb shift, which will extensively discussed in the next chapters. Many authors have discussed and commented Bethe's nonrelativistic calculations for the Lamb shift. P.A.M. Dirac was the first to argue that intelligent mathematics consists of neglecting negligible quantities and not infinite quantities because, simply, you don't want them. In fact, the original method that Bethe introduced was to calculate finite quantities by subtracting and neglecting infinite quantities. However, as many authors remarked, what is extremely puzzling is that after manipulating and ignoring infinite quantities, after imposing arbitrary integration limits, after considering mean logarithmic values over an infinity of atomic energy levels near the continuum, "the final result compares "remarkably well" to the experiment".

We have not presented in this chapter, the relativistic calculation of the Lamb shift, developed and published in 1949 by Kroll *et al.* because the calculations are extremely tedious. Nevertheless, we have to mention that even in this approach various approximations are introduced and a "different" mean logarithmic value is employed. Furthermore, it seems that the upper integration limit appears naturally in the relativistic approach, however, the author, as the majority of the authors in the literature, has not attempted to reproduce completely these calculations.

Bibliography

1. A.I. Akhiezer and B.V. Berestetskii, *Quantum electrodynamics*, New York: Interscience Publishers. 1965.
2. Y. Aharonov and D. Bohm, Significance of electromagnetic potentials in the quantum theory, *Phys. Rev.* **115**(3) (1959) 485–491.
3. G. Auletta, *Foundations and interpretation of quantum mechanics*, Singapore: World Scientific. 2001.
4. G. Beck, Zur theorie des photoeffekts, *Z. Phy.* **41** (1927) 443–452.
5. H.A. Bethe, The electromagnetic shift of energy levels, *Phys. Rev.* **72**(4) (1947), 339–341.
6. N. Bohr, H.A. Kramers and J.C. Slater, The quantum theory of radiation, *Phil. Mag.* **47**(281) (1924) 785–802.
7. W. Bothe and H. Geiger, Über das wesen des compton effekts: ein experiment ellerbeitrag zur theories der strahlung, *Z. Phys.* **32**(9) (1925) 639–663.
8. B.H. Bransden and C.J. Joachain, *Physics of atoms and molecules*, London: Longman Group Ltd. 1983.

9. R.G. Chambers, Shift of an electron interference pattern by enclosed magnetic flux, *Phys. Rev. Lett.* **5**(1) (1960) 3–5.
10. S.L. Chuang, *Physics of photonic devices,* New Jersey: John Wiley & Sons. 2009.
11. A.H. Compton, The spectrum of scattered x-rays, *Phys. Rev.* **22**(5) (1923) 409–413.
12. L. de Broglie, *The revolution in physics; a non-mathematical survey of quanta,* New York: Noonday Press. 1953, pp. 117, 178–186.
13. P.A.M. Dirac, *The principles of quantum mechanics,* Oxford: Oxford University Press. 1958.
14. W. Ehrenberg and R.E. Siday, *Proceedings of the Physical Society B* **62**, 1949, pp. 8–21.
15. R. Feynman, *The strange theory of light and matter,* Princeton, New Jersey: Princeton University Press. 1988.
16. J.C. Garrison and R.Y. Chiao, *Quantum optics,* Oxford: Oxford University Press. 2008.
17. H. Haken, *Light,* Amsterdam–Oxford: North Holland Publishing. 1981.
18. W. Heitler, *The quantum theory of radiation,* Oxford: Clarendon Press. 1954.
19. N.M. Kroll and W.E. Lamb Jr, On the self-energy of a bound electron, *Phys. Rev.* **75**(3) (1949) 388.
20. W.E. Lamb Jr and M.O. Scully, In *Polarization, matière et rayonnement, volume jubilaire en l' honneur d' Alfred Kastler,* French Physical Society, Paris: Press Universitaires de France. 1969.
21. J.-M. Liu, *Photonic devices,* Cambridge: Cambridge University Press. 2005.
22. P.W. Milonni, *The quantum vacuum,* London: Academic Press Inc. 1994.
23. M.H. Mittleman, *Introduction to the theory of laser-atom interactions,* New York: Plenum Press. 1982.
24. M. Planck, *The theory of heat radiation,* New York: Dover Publications. 1959.
25. L.H. Ryder, *Quantum field theory,* Cambridge: Cambridge University Press. 1987.
26. B.E.A. Saleh and M.C. Teich, *Fundamentals of photonics,* New York: John Wiley & Sons. 2007.
27. R.E. Siday, The optical properties of axially symmetric magnetic prisms, *Proc. Phys. Soc.* **59**(6) (1947) 1036.
28. G.I. Taylor, Interference fringes with feeble light, *Proc. Camb. Philos. Soc.* **15** (1909) 114–115.
29. M. Weissbluth, *Photon-atom interactions,* New York: Academic Press Inc. 1988.
30. G. Wentzel, *Z. Phys.* **40** (1926) 574.

Chapter 5

Theory, Experiments and Questions

In what follows, we analyze the questions raised from the quantum theory of radiation and we point out the mathematical difficulties encountered mainly from the second quantization procedure which has been developed to provide photons description from the quantum point of view. We first make a theoretical analysis of some particular aspects of QED, including the well known singularities, and then we present some experiments which strongly support the simultaneous wave-particle nature of photons.

5.1. Planck's Constant and the Vacuum Intrinsic Electromagnetic Properties

The correspondence for the energy between the electromagnetic wave theory and the Quantum electrodynamics (QED) describing the same entity, the light, is given by the well-known expression

$$E\left(\omega\right) = \hbar\omega \leftrightarrow E\left(\omega, A_0(\omega), \varepsilon_0\right) = \int_V 2\varepsilon_0\omega^2 A_0^2(\omega, \vec{r})\, d^3r \qquad (5.1.1)$$

In QED, the energy of the photon $\hbar\omega$ depends only on the angular frequency of the electromagnetic wave ω and it is proportional to Planck's reduced constant $\hbar$, while the vacuum electric permittivity constant ε_0 and the vector potential amplitude $A_0(\omega)$, which both play a major role in the classical electromagnetic wave description (3.4.8), are totally ignored. On the other hand, it is obvious that the knowledge of the angular frequency alone is not sufficient in the classical electromagnetic theory in order to deduce the electromagnetic field energy. The vector potential amplitude is indispensable.

Consequently, for (5.1.1) to be physically coherent, $\hbar$ should be related to both the vacuum electric permittivity ε_0 and the vector potential amplitude $A_0(\omega)$.

Indeed, $\hbar$ is related to ε_0 through the fine structure constant α_{FS} (4.6.4) according to the relation:

$$\hbar = \left(\frac{e^2}{4\pi\varepsilon_0}\right)\left(\frac{1}{c\alpha_{FS}}\right) \qquad (5.1.2)$$

where e is the electron charge.

From the last equation one can also express in QED, the energy of a photon depending on ε_0 as follows

$$E(\omega) = \hbar\omega = \left(\frac{e^2}{4\pi\varepsilon_0}\right)\left(\frac{1}{c\alpha_{FS}}\right)\omega \qquad (5.1.3)$$

According to (5.1.3), one may assume that the absolute value of the energy of a photon with an angular frequency ω corresponds to the Coulomb interaction between an electron and a positron separated by the characteristic distance l_ω

$$l_\omega = \frac{c\alpha_{FS}}{\omega} = \alpha_{FS}\frac{\lambda}{2\pi} \qquad (5.1.4)$$

where λ is the wavelength.

However, it is puzzling to note that the expression (5.1.3), introduces the charge of the electron and the fine structure constant which essentially characterize the atomic energy levels. Consequently, one could naturally wonder why these physical quantities should be necessary to describe the energy of light in vacuum, in complete absence of charges and matter? In fact, the electron (positron) and the electromagnetic field are strongly related physical entities. Under particular conditions, in the presence of a strong electric field like in the vicinity of a heavy nucleus, a high energy gamma photon can be annihilated, giving birth to an electron and a positron. Of course, the probability of such a phenomenon is extremely low and becomes significant for a nucleus of $Z \sim 140$, which is physically instable. Nevertheless, Fulcher *et al.* have argued that such a nuclear state could be created during the collision of two nuclei of large Z. Thus, we can say that (5.1.3) is a simple relation which implies directly that the electron (positron) and the electromagnetic waves are physical entities that might be issued from the same quantum field related to vacuum.

Now, we have seen that we can easily establish a relation between $\hbar$ and the vacuum permittivity ε_0 through the fine structure constant α_{FS} but what would be the relation between $\hbar$ and the vector potential amplitude $A_0(\omega)$ for a single photon guessed by (5.1.1)? In a coherent description of the photon as a quantum of the electromagnetic field, this relation should be independent of any external parameter such as the volume V involved in the link equations (4.3.14).

We will explore the possibility of such a relation in the next chapter.

5.2. Hamiltonian Issued from the Quantization of the Electromagnetic Field

Let us go back to the quantization procedure of the electromagnetic field. When we introduce the link expressions (4.3.14), directly in the relation of the energy of the classical electromagnetic field (4.3.9), we obtain the following radiation Hamiltonians

$$H_{\mathrm{EM}} = \sum_{k,\lambda} \hbar \omega_k \hat{N}_{k\lambda} \qquad (5.2.1)$$

if the "normal ordering" expression is considered in (4.3.9), i.e.,

$$2\varepsilon_0 V \sum_{k,\lambda} \omega_k^2 A_{k\lambda}^* A_{k\lambda} \qquad (5.2.2)$$

and

$$H_{\mathrm{EM}} = \sum_{k,\lambda} \hbar \omega_k (\hat{N}_{k\lambda} + 1) \qquad (5.2.3)$$

when the anti-normal ordering is used

$$2\varepsilon_0 V \sum_{k,\lambda} \omega_k^2 A_{k\lambda} A_{k\lambda}^* \qquad (5.2.4)$$

It is very important to note that it is impossible to get $H_{\mathrm{EM}} = \sum_{k,\lambda} \hbar \omega_k (\hat{N}_{k\lambda} + \frac{1}{2})$ when starting from the electromagnetic field energy expression (4.3.9), the "normal ordering" is used or not.

This mathematical ambiguity, which leads to a quite different physical interpretation of the radiation zero-level state, demonstrates the lack of a coherent mathematical transition from the classical to the quantum mechanical formalism of the electromagnetic field energy.

Consequently, a strong doubt may arise in whether the term $\sum_{k,\lambda} \frac{1}{2} \hbar \omega_k$, corresponding to the vacuum energy in (4.3.18) represents a real physical state since it is not in reality a natural consequence of the electromagnetic field quantization. Hence, we will analyze in next chapters, the reasons for which one obtains different expressions for the zero-point energy. Also we will show how the relation (4.3.12) has to be transformed to ensure a coherent transition between the classical electromagnetic wave energy and the quantum mechanical Hamiltonian, in order to eliminate this mathematical ambiguity.

5.3. QED Singularities

Let us consider the Hamiltonian of the electromagnetic field (4.3.18) in complete absence of photons, that is when $N_{k,\lambda} = 0$, which becomes

$$H_{\text{vacuum}} = \sum_{k,\lambda} \frac{1}{2} \hbar \omega_k \qquad (5.3.1)$$

The summation runs over all modes k and polarizations λ. Hence, as it has been mentioned in Chapter 4 (4.3.1), the vacuum in the QED description consists of an infinite sea of "photons" corresponding to the state $|N_{k,\lambda} = 0, \forall \{k, \lambda\}\rangle$ corresponding to infinite energy.

In this context, the vacuum electromagnetic energy density is

$$W_{\text{vacuum}} = \frac{1}{V} \sum_{k,\lambda} \frac{1}{2} \hbar \omega_k \qquad (5.3.2)$$

using (4.6.2) and considering two polarizations, we get

$$W_{\text{vacuum}} = \frac{\hbar}{2\pi^2 c^3} \int \omega^3 d\omega \qquad (5.3.3)$$

which is also infinite.

In the visible region situated in the wavelengths range between 4000 and 7000 Angstroms, the energy density calculated using (5.3.3) corresponds roughly to $22\,\text{J}/\text{m}^3$, which represents a tremendous amount of energy in space which is fully unobservable experimentally. Recent astronomical observations have demonstrated that the measured vacuum energy density in space is many orders of magnitude less than that predicted by QED theory. The huge discrepancy between the theoretical and the measured values of the vacuum energy has been called the "quantum vacuum catastrophe."

On the other hand, the quantized radiation electric field $\vec{E}(\vec{r}, t)$ is obtained using the vector potential expression (4.3.15) and (3.3.15) in the Coulomb gauge

$$\vec{E}(\vec{r}, t) = -\frac{\partial \vec{A}(\vec{r}, t)}{\partial t}$$

$$= i \sum_{k,\lambda} \sqrt{\hbar \omega_k / (2\varepsilon_0 V)}$$

$$\times [a_{k\lambda} \hat{\varepsilon}_{k\lambda} e^{i(\vec{k}\cdot\vec{r} - \omega_k t + \varphi)} - a_{k\lambda}^+ \hat{\varepsilon}_{k\lambda}^* e^{-i(\vec{k}\cdot\vec{r} - \omega_k t + \varphi)}]$$

$$(5.3.4)$$

where $\hat{\varepsilon}_{k\lambda}$ is a polarization vector. Consequently, the expectation value of the photons electric field in vacuum within a volume V is given by the expression

$$\langle E(\vec{r})^2 \rangle_{\text{vacuum}} = \sum_{k,\lambda} \frac{\hbar \omega_k,}{2\varepsilon_0 V} \qquad (5.3.5)$$

which is also infinite.

5.4. Electron-Vacuum Interactions and the Associated Effects

The presence of the vacuum energy singularity in the quantised radiation field Hamiltonian $H_{\text{vacuum}} = \sum_{k,\lambda} \frac{1}{2}\hbar \omega_k$ constitutes the fundamental argument generally found in the literature, in order to explain some physical effects such as the spontaneous emission, Lamb displacements in atomic levels as well as the Casimir effect. The terminology usually employed for that purpose uses the term "vacuum fluctuations", though in most cases no mathematical representation of any kind of fluctuation is explicitly presented. Some authors prefer using the term "virtual photons" for the vacuum Hamiltonian.

However, we could naturally wonder why the first term of the Hamiltonian (4.3.18) is considered as real while the second one as "virtual".

Nevertheless, it is absolutely certain that the last effects are due to the electron-vacuum interactions and cannot be calculated using a semi-classical model, in which the atomic energy levels are quantized while the electric field is represented by a classical wave equation in the interaction Hamiltonian. The reason is simply that in the classical description, the vacuum has no electromagnetic waves and consequently, $\vec{E}(\vec{r}, t)$ in the interaction Hamiltonian H_{int} in (4.5.26) is zero. Conversely, in QED, according to the radiation Hamiltonian (4.3.18), the vacuum is considered to be composed by photons of all modes and polarizations, permitting one to consider an interaction Hamiltonian.

Now, the above effects are due to the interaction between an electron in a given atomic state and the vacuum. Consequently, for their description the term $\sum_{k,\lambda} \frac{1}{2} \hbar \omega_k$ of the Hamiltonian (4.3.18), which is the eigenvalue of the fundamental radiation eigenstate $|\ldots, 0_{k,\lambda}, \ldots, 0_{k',\lambda'}, \ldots\rangle$, corresponding to the vacuum state, should be of high importance.

However, as we will see in the following, this singularity has no implication in the mathematical interpretation of those effects.

• Spontaneous emission

We recall the total Hamiltonian of an atom in the presence of an electromagnetic field in the dipole approximation:

$$H_{\text{tot}} = H_{\text{EM}} + H_{\text{int}} + H_{\text{atom}}$$

$$= \sum_{k,\lambda} \hbar \omega_k \left(\hat{N}_{k\lambda} + \frac{1}{2} \right) - \vec{D} \cdot \vec{E}(\vec{r}, t) + \sum_i \hbar \omega_i |\Psi_i\rangle\langle\Psi_i|$$

$$(5.4.1)$$

where $\vec{D} = e\vec{r}$ is the dipole moment of an atomic electron of charge e according to (4.5.26) and $|\Psi_i\rangle$ the atomic levels with the corresponding energies $\hbar \omega_i$.

The spontaneous emission rate is always calculated in QED neglecting $\sum_{k,\lambda} \frac{1}{2} \hbar \omega_k$ in H_{tot} and using the expression (5.3.4) for the electric field $\vec{E}(\vec{r}, t)$.

Therein, we recall that according to Heisenberg's equation of motion (4.5.22.a), the electric field writes

$$\vec{E}(\vec{r},t) = -\frac{\partial \vec{A}(\vec{r},t)}{\partial t} = \frac{i}{\hbar}\left[\vec{A}(\vec{r},t), \sum_{k,\lambda} \hbar\omega_k \left(\hat{N}_{k\lambda} + \frac{1}{2}\right)\right] \qquad (5.4.2)$$

Consequently, for the calculation of the electric field attributed to the vacuum state, the term $\sum_{k,\lambda} \frac{1}{2}\hbar\omega_k$ should be the principal contribution to the Hamiltonian. But the commutation operator of the vector potential and the vacuum Hamiltonian cancels

$$\left[\vec{A}(\vec{r},t), \sum_{k,\lambda} \frac{1}{2}\hbar\omega_k\right] = 0 \qquad (5.4.3)$$

Hence, the term $\sum_{k,\lambda} \frac{1}{2}\hbar\omega_k$ in this representation has absolutely no contribution in the general expression of the quantized electric field used to calculate the spontaneous emission.

Obviously, the vacuum effect in the quantized vector potential is involved in the commutation relation $[a_{k\lambda}, a_{k\lambda}^+] = 1$ and not in the $\sum_{k,\lambda} \frac{1}{2}\hbar\omega_k$ term.

Indeed, in a rigorous calculation, the interaction Hamiltonian between an atomic electron and the vacuum writes

$$H_{\text{int}} = -\vec{d} \cdot \frac{i}{\hbar}\left[\vec{A}(\vec{r},t), \sum_{k,\lambda} \frac{1}{2}\hbar\omega_k\right] = 0 \qquad (5.4.4)$$

which also vanishes in QED, as in the semi-classical description.

The physical reason is that in QED, the vacuum state Hamiltonian is not described by a function of $a_{k\lambda}$ and $a_{k\lambda}^+$ operators and consequently it is impossible to define an interaction Hamiltonian for the description of the electron-vacuum interaction processes.

• Lamb shift

The mathematical treatment of the quantum vacuum effects upon the atomic energy levels described in Section 4.6 is without any doubt among the most revolutionary ones introduced in modern physics. It consists of neglecting infinite quantities for calculating finite energy differences which are measurable experimentally.

We can now have a close look to a detailed analysis of this calculation, resuming the comments and arguments of many authors.

First, the quadratic term of the vector potential of equation
(4.6.1) is dropped out, though it's infinite. Despite the fact that the
linear term of the vector potential of equation (4.6.3) is also infinite,
it is not dropped out as the quadratic one, but instead it suffers a
renormalization procedure which equally results in an infinite quantity (4.6.6). At that level one should expect that since the quadratic
term, being infinite, does not induce a measurable shift, by the same
token the linear term, being also infinite, should also not induce a
measurable shift. Nonetheless, the linear term is kept in the calculation and in order to "overcome" the difficulty, an arbitrary upper
limit for the integration of (4.6.6) is introduced in the nonrelativistic
approach whose choice, at that level of calculation, is not based on
any experimental or physical argument. Conversely, in the relativistic
approach developed much later, the upper integration limit seems to
appear "naturally". Consequently, one may wonder why the same
upper limit is not equally applied in the integration of the quadratic
term of the vector potential. Furthermore, the introduction of a mean
value for the logarithmic term (4.6.9), calculated numerically over
an indefinite number of discrete states, in both the nonrelativistic
and the relativistic approach, gives rise to the upsetting feeling that
under these calculation conditions, the fact that the final obtained
displacement energy is accurate compared to the experiment is
puzzling.

Finally, once again, in these calculations the vacuum effects on
the atomic energy levels is not due to the zero-point level of the
Hamiltonian (4.3.18). Except Bethe's approach, various calculations
of the Lamb shift have been published, each one by considering the
influence of the vacuum upon the atomic orbitals through different
aspects; Welton's vacuum electric field perturbation, Feynman's gas
interpretation, Stark shift due to the vacuum electric field and so
forth. It is quite interesting noticing that almost all these physical
mechanisms used for the interpretation of the atomic energy shifts
do no require any mass renormalization process.

• Casimir effect

The behavior of two perfectly conducting and uncharged parallel plates has been investigated by Casimir in 1948. This physical
situation has been extensively studied by many authors over half a
century and of course it is out of the scope of this book to present
all these works in detail. Considering the energy difference of the
vacuum radiation modes between and outside the plates, separated
by a distance d, it is drawn out that an attractive force arises, whose
expression is

$$F(d) \approx -(4.11 \ 10^{-2})\frac{\hbar c}{d^4} \qquad (5.4.5)$$

The explanation initially advanced, and which is still generally
admitted today, is that when considering the boundary conditions,
as in (3.5.39), the density of the vacuum photons outside the plates
is higher than that in the region inside, inducing a radiation pressure
pushing the plates together to come closer.

The first experiments carried out by Sparnaay in 1958 and then by
van Blokland and Overbeek in 1978, in order to measure Casimir's
force were not quite successful, since the experimental errors were
very significant, nearly 100% and 30% respectively. However, they
permitted to put in evidence an attractive force between the plates.

More recent experiments mentioned by Lambrecht and Reynaud
in 2002, attained a precision of the order of 1%, but they were still
unable of confirming neither the theoretical predictions at larger distances (of the order of a μm) nor the temperature effects. Hence,
it appears that higher accuracy experiments are further required to
fully confirm the agreement with Casimir's theory.

Despite this, the Casimir effect is generally considered as the principle demonstration of the existence of the vacuum zero-point level
photons $\sum_{k,\lambda} \frac{1}{2}\hbar\omega_k$ involved in the Hamiltonian (4.3.18). However,
it is important underlying that Schwinger *et al.* in 1978 and Milonni
in 1982 obtained the same expression for the attraction force with
that of Casimir's, without even referring to the zero-point energy. In
fact, as mentioned by Milonni in 1992, a simple summation of the

Van der Waals pairwise intermolecular forces on the surfaces of the two plates gives a quite close result to that of the Casimir force

$$F(d) \approx -(3.27 \; 10^{-2})\frac{\hbar c}{d^4} \qquad (5.4.6)$$

demonstrating that physical and reasonable results can be obtained from classical electrodynamics, without considering the hypothetical energy density difference of the vacuum photons in the regions between and outside the plates. As a conclusion, caution has to be taken concerning the interpretation of the physical origin of Casimir's effect since Casimir's theory for larger distances and the associated temperature effects have not yet been confirmed experimentally.

5.5. Simultaneous Wave-particle Nature of the Photon Revealed by the Experiments — Discussions

• Photoelectric effect and Young's double-slit experiment

The photoelectric effect was initially considered as a direct demonstration of the particle nature of light and historically was the strongest argument of the photon concept, although Einstein had not advanced any explanation on what could correspond the frequency ν for a point particle.

Nevertheless, as we have seen in Chapter 4 (4.5), Wentzel in 1926 and Beck in 1927, as well as much later, Mendel, Lamb and Scully in the 1960s, demonstrated that the photoelectric effect can be interpreted remarkably well by considering the electromagnetic wave nature of the light issued from Maxwell's equations without referring at all to the photon concept.

Consequently, since both wave and particle models interpret satisfactorily the photoelectric effect, it cannot be considered as a decisive experiment for the introduction of the photon concept.

On the other side, Young's double-slit experiment, which was initially interpreted with the wave nature of light and was used as the strongest argument against the particle theory, was used by Taylor at very low intensities to demonstrate the particle concept.

The interpretation of Taylor's experiment by some authors was based on the wave nature of a single photon permitting to pass through both slits. Many scientists were skeptical about this interpretation, as in recent similar experiments, interference patterns have been observed with heavy atoms and molecules. Indeed, Nairz *et al.* obtained recently interferences with spherical fullerenes, which are almost six orders of magnitude bigger than a nucleon particle. It is impossible to conceive, at least in the present stage of our knowledge that fullerenes can pass through both slits simultaneously.

In order to get a comprehensive picture of all these interference experiments carried out with light and material particles, Jin *et al.* in 2010 advanced a particle-based description of Young's double slit interferences without even referring to the wave theory.

Consequently, what can be essentially deduced from the experimental evidence during more than two centuries of struggling between the wave and particle theories, is that experiments simply confirm both natures for light (and for material particles).

Bohr, although initially in favor of the wave concept, understood quite rapidly that the only way out of this frustrating dilemma is the wave-particle complementarity, even if this concept is hardly comprehensive for the human common sense.

We will now present selected recent experiments and discuss the final results with respect to the second quantization procedure.

- Mizobuchi and Ohtake double-prism experiment

Since Bohr announced his famous complementarity principle expressing that light has simultaneously both the wave and particle nature, appearing mutually exclusive in every physical situation, the opposition between the defenders of the wave or particle theories had ceased. However, the majority of scientists were skeptical about the wave-particle simultaneity which is not at all comprehensive for the human mind.

With the development of lasers and the revolutionary parametric down converters techniques, it becomes possible to realize conditions in which, with a very high degree of statistical accuracy, only one photon is present in the experimental apparatus. Hence, in order to test Bohr's complementarity principle, Ghose, Home and Agarwal, proposed in 1991, the double prism experiment which was carried out by Mizobuchi and Ohtake in 1992.

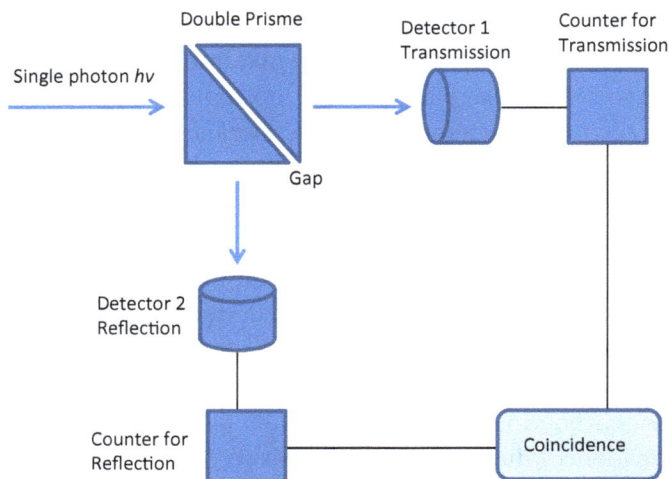

Fig. 5.5.1. Schema of the double prism experiment initially proposed by Ghose, Home and Agarwal and carried out by Mizobuchi and Ohtake. With kind permission from Springer Science+Business Media: <Foundations of Physics, The Two-Prism Experiment and Wave-Particle Duality of Light, **26**(7), 943–953, Fig. 1 1996, Partha Ghose and Dipankar Home, and any original (first) copyright notice displayed with material>.

A single photon pulse goes through two prisms separated by an air gap whose thickness is less than the incident photon wavelength (Fig. 5.5.1). In these conditions, classical electromagnetic wave theory predicts a tunneling effect for the wave-like photon, according to which the wave will traverse the air gap straightforward and will be detected by Detector 1. On the other hand, a particle-like photon will be reflected by the inner wall of the first prism and will be detected by Detector 2. Both detectors are connected to a coincidence counter in order to detect any eventual coincidence.

The experiment results have shown very strong anti-coincidence entailing that a single photon state goes either through the gap by wave tunneling or is reflected by the first prism like a particle. Consequently, both the wave and the particle natures of the photon are exhibited in the same experimental conditions, confirming the wave-particle concept and at the same time contradicting the mutual exclusiveness of Bohr's complementarity principle.

Finally, it is worth noting that simultaneous wave-particle duality for neutrons in similar interferometer have also been reported by Greenberger and Yasin.

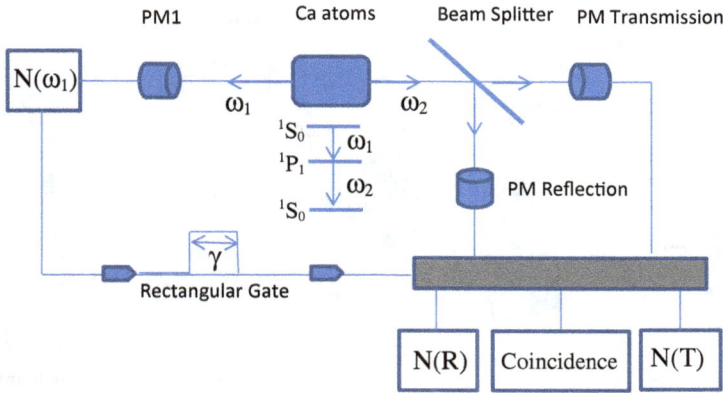

Fig. 5.5.2. Experimental disposition of Grangier's *et al.* experiment for the demonstration of the indivisibility of a single photon state. Reproduced with permission from Ref. 19.

- Grangier's *et al.* experiments, photon indivisibility

The purpose of this experiment is to test the indivisibility of the photon. It is based on the two emission lines of the lower energy levels of Ca atoms $^1S_0 \rightarrow^1 P_1$ and $^1P_1 \rightarrow^1 S_0$ (Fig. 5.5.2). The first transition gives a photon with angular frequency ω_1, whose detection opens a rectangular gate to control the detection of the second photon emission at ω_2. The last one passes through a beam splitter behind which photomultipliers can detect the transmitted and the reflected ω_2 photons. The intermediate level has a life time τ and in order to decrease the probability to get coincidence, counting due to two ω_2 photons issued from two different Ca atoms, the gate width γ has been chosen to be much smaller than τ.

The final results of the experiment revealed extremely low levels of coincidences, demonstrating without any doubt, that the photons are integral particles transmitted or reflected by the beam splitter as a whole.

Although the issue of the experiment shows the indivisibility of the photons, it does not exclude the wave nature of them.

- Hunter–Wadlinger experiments, the photon spatial expansion

Akhiezer and Berestetskii have pointed out that it is impossible to consider a photon within a length shorter than the wavelength.

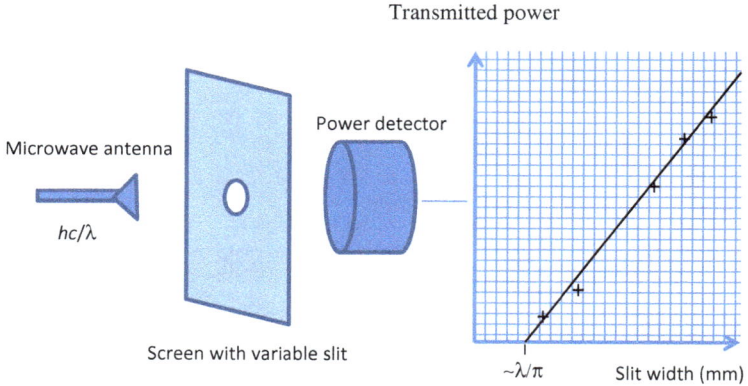

Fig. 5.5.3. Schema of Hunter's and Wadlinger's experiment for the investigation of the lateral spatial extension of the electromagnetic waves. Reproduced with permission from Ref. 21.

Nowadays, this concept is generally accepted. However, the lateral expansion of the photon considered as an integral particle was always an intriguing part of physics.

For many decades, it has been well known that Faraday's metallic cage is a quite efficient shield toward electromagnetic waves, provided that the dimensions of the holes of the grid are roughly five times smaller than the wavelength of the incident radiation, which is $\sim \lambda/5$. That was rather an empirical rule and most of the constructed grills to screen the radiation of a wavelength λ imposed the dimensions of the holes to be within the interval $\lambda/4$ to $\lambda/8$.

Robinson in 1953 and Hadlock in 1958 carried out experiments using microwaves crossing small apertures, and deduced that no energy is transmitted through apertures whose diameters are less than roughly $\sim \lambda/4$.

In 1986, Hunter and Wadlinger, realized a very simple experiment inspired by those carried out by Robinson and Hadlock (Fig. 5.5.3). They used x-band microwaves with $\lambda = 28.5\,\mathrm{mm}$ and measured the transmitted power through rectangular or circular apertures of different dimensions.

The purpose was to define the smaller dimensions of the slit beyond which the transmitted power is zero. All the results, corrected with the subtraction of the first harmonic of the emitted beam by the microwave antenna, have shown that the cut-off dimension (diameter for the circular aperture, width for the rectangular one) is roughly $\sim \lambda/\pi$.

Consequently, taking also into account that the longitudinal extension of the photon along the propagation axis is the wavelength λ itself, it implies that the intrinsic "photon volume" depends on its wavelength and it should be proportional to λ^3.

5.6. Conclusion Remarks

We have seen that the equivalence of the photons energy with that of the classical electromagnetic wave description implies that a relationship between Planck's constant and the vector potential quantized amplitude should exist. We have also analyzed the QED singularities and shown that the term $\sum_{k,\lambda} \frac{1}{2}\hbar\omega_k$ is not involved in the vacuum-electron interactions.

On the experimental front, we may essentially deduce that the photoelectric effect and Young's interferences have both demonstrated the wave-particle nature of light. This has been further established by Mizobuchi and Ohtake's experiment, which also demonstrates that the wave and particle natures appear simultaneously and are not necessarily exclusive.

On the other hand, Robinson, Hunter and Wadlinger have demonstrated that the photon should have a lateral extension of the order $\lambda/4$ to λ/π and cannot be a "point particle," while the experiments of Grangier *et al.* demonstrated the photon indivisibility.

During the Compton scattering experiments, the photon exhibits a very particular property; it conserves its integrity as particle, but may concede a fraction of its energy, entailing a modification of its wavelength and consequently its energy, momentum and spatial extension.

To sum up the experimental evidence, it turns out that the photon appears to be an indivisible quantum (segment) of the electromagnetic field over a period, thus extended along a wavelength, with intrinsic wave properties, emitted and detected as a whole and capable of interacting with charged particles, increasing (or even decreasing) its wavelength and consequently decreasing (or even increasing) its energy.

In the next chapter, we will consider further theoretical elaborations conform to the experiments and advance a non-local photon representation based on a precise quantization of the electromagnetic field vector potential, linking coherently the electromagnetic theory and quantum mechanics and entailing a well-defined description of the quantum vacuum.

Bibliography

1. G.S. Agarwal, Quantum electrodynamics in the presence of dielectrics and conductors. II. Theory of dispersion forces, *Phys.Rev. A* **11** (1975) 243.
2. A.I. Akhiezer and B.V. Berestetskii, *Quantum electrodynamics*, New York: Interscience Publishers. 1965.
3. C.L. Andrews, *Optics of the electromagnetic spectrum*, Englewood cliffs, New Jersey: Prentice-Hall Inc. 1960.
4. A. Aspect, *The wave-particle dualism*, S. Diner, D. Fargue, G. Lochak, & F. Selleri (Eds.) Dordrecht: D. Reidel Publication Company. 1984, pp. 297–330.
5. G. Auletta, *Foundations and interpretation of quantum mechanics*, Singapore: World Scientific. 2001.
6. G. Beck, Zur theorie des photoeffekts, *Z. Phys.* **41** (1927) 443–452.
7. H.A. Bethe, The electromagnetic shift of energy levels, *Phys. Rev.* **72**(4) (1947), 339–341.
8. N. Bohr, H.A. Kramers and J.C. Slater, The quantum theory of radiation, *Phil. Mag.* **47**(281) (1924) 785–802.
9. S. Bourzeix, B. de Beauvoir, F. Bez, M.D. Plimmer, F. de Tomasi, L. Julien, F. Birabenand D.N. Stacey, High resolution spectroscopy of the hydrogen atom: determination of the 1S Lamb shift, *Phys. Rev. Lett.* **76** (1996) 384.
10. W. Bothe and H. Geiger, Über das wesen des compton effekts: ein experiment ellerbeitrag zur theories der strahlung, *Z. Phys.* **32**(9) (1925) 639–663.
11. B.H. Bransden and C.J. Joachain, *Physics of atoms and molecules*, London: Longman Group Ltd. 1983.
12. H.B.G. Casimir and D. Polder, The influence of retardation on the London-van der Waals forces, *Phys. Rev.* 73 (1948) 360.
13. A.H. Compton, The spectrum of scattered x-rays, *Phys. Rev.* **22**(5) (1923) 409–413.
14. A. Einstein, *The collected papers of Albert Einstein*, J. Stachel, D.C. Cassidy, J. Renn, & R. Schulmann (Eds.) Princeton, New Jersey: Prinston University Press. 1987.
15. R. Feynman, *The strange theory of light and matter*, Princeton, New Jersey: Princeton University Press. 1988.
16. J.C. Garrison and R.Y. Chiao, *Quantum optics*, Oxford: Oxford University Press. 2008.

17. P. Ghose and D. Home, The two-prism experiment and wave-particle. Duality of light, *Found. Phys.* **26**(7) (1996) 943–953.
18. D.M. Greenberger and A. Yasin, Simultaneous wave and particle knowledge in a neutron interferometer, *Phys. Lett. A* **128** (1988) 391–394.
19. P. Grangier, G. Roger and A. Aspect, *Europhysics Lett.*, **1**(4) (1986) 173–179. http://iopscience.iop.org/0295-5075
20. R.K. Hadlock, Diffraction patterns at the plane of a slit in a reflecting screen, *J. Appl. Phys.* **29**(6) (1958) 918–920.
21. G. Hunter and R.L.P. Wadlinger, *Phys. Essays* **2**(2) (1989) 158.
22. G. Hunter, "Quantum uncertainties" NATO ASI Series B, *Proceedings, NATO Advanced Research Workshop*, 1986 June 23–27, pp. 331.
23. E.T. Jaynes and F.W. Cummings, Comparison of quantum and semiclassical radiation theories with application to the beam maser, *Proc. IEEE* **51**(1) (1963) 89–109.
24. S. Jeffers, S. Roy, J.P. Vigier and G. Hunter, *The present status of the quantum theory of light*, Boston: Kluwer Academic Publishers. 1997.
25. F. Jin, S. Yuan, H. De Raedt, K. Michielsen and S. Miyashita, Corpuscular model of two-beam interference and double-slit experiments with single photons, *J. Phys. Soc. Jpn.* **79**(7) (2010) 074401, DOI: org/10.1143/JPSJ.2010. 79.074401.
26. W.E. Lamb Jr and M.O. Scully, In *Polarization, matière et rayonnement, volume jubilaire en l'honneur d'Alfred Kastler*, French Physical Society, Paris: Press Universitaires de France. 1969.
27. A. Lambrecht and Reynaud S., In *Séminaire Poincaré 2002: vacuum energy-renormalization, Prog Math Phys 30*, V. Rivasseau, & B. Duplantier (Eds.) Birkhäuser, Basel. 2003, pp. 109.
28. S.K. Lamoreaux, Demonstration of the Casimir force in the 0.6 to 6 μm range, *Phys. Rev. Lett.* **78**(1) (1997) 5.
29. C. Meis, Electric potential of the quantum vacuum, *Phys Essays* **12**(1) (1999).
30. P.W. Milonni, *The quantum vacuum*, London: Academic Press Inc. 1994.
31. P.W. Milonni, Casimir forces without the vacuum radiation field, *Phys. Rev. A* **25**(3) (1982) 1315.
32. P.W. Milonni and M.L. Shih, Source theory of the Casimir force, *Phys. Rev. A* **45**(7) (1992) 4241.
33. U. Mohideen and A. Roy, Precision measurement of the Casimir force from 0.1 to 0.9 μm, *Phys. Rev. Lett.* **81**(21) (1998) 4549.
34. O. Nairz, M. Arndt and A. Zeilinger, Quantum interference experiments with large molecules, *Am. J. Phys.* **71**(4) (2003) 319–325.
35. O. Nairz, B. Brezger, M. Arndt and A. Zeilinger, Diffraction of complex molecules by structures made of light, *Phys. Rev. Lett.* **87** (2001) 160401, DOI: org/10.1103/PhysRevLett.87.160401.
36. N. Osakabe, T. Matsuda, T. Kawasaki, J. Endo, A. Tonomura, S. Yano and H. Yamada, Experimental confirmation of Aharonov-Bohm effect using a toroidal magnetic field confined by a superconductor, *Phys. Rev. A* **34**(2) (1986) 815.

37. R.L. Pfleegor and L. Mandel, Interference of independent photon beams, *Phys. Rev.* **159**(5) (1967) 1084–1088.
38. H.L. Robinson, Diffraction patterns in circular apertures less than one wavelength in diameter, *J. Appl. Phys.* **24**(1) (1953) 35.
39. A. Roy and U. Mohideen, Demonstration of the non trivial boundary dependence of the Casimir force, *Phys. Rev. Lett.* **82**(22) (1999) 4380.
40. J. Schwinger, L.L. DeRaad Jr. and K.A. Milton, Casimir effect in dielectrics, *Ann. Phys. (N.Y.)* **115**(1) (1978).
41. M.J. Sparnaay, Measurements of attractive forces between flat plates, *Physica* **24**(6–10) (1958) 751–764.
42. G.I. Taylor, Interference fringes with feeble light, *Proc. Camb. Philos. Soc.* **15** (1909) 114–115.
43. P.H.G.M. van Blokland and J.T.G. Overbeek, Van der Waals forces between objects covered with a chromium layer, *J. Chem. Soc. Faraday Trans.* **74** (1978) 2637–2651.
44. T.A. Welton, Some Observable Effects of the Quantum-Mechanical Fluctuations of the Electromagnetic Field, *Phys. Rev.* **74**(9) (1948) 1157.
45. G. Wentzel, Zur theorie des photoelektrischen effekts, *Z. Phys.* **40** (1926) 574–589.

Chapter 6

Analysis of the Electromagnetic Field Quantization Process and the Photon Vector Potential. The Non-Local Photon Wave-Particle Representation and the Quantum Vacuum

6.1. Quantized Vector Potential Amplitude of a Single Photon State

Frequency and vector potential imply electromagnetic wave properties while energy and momentum imply particle properties. In a full wave-particle description of light, the challenge consists of establishing a coherent relation among all these physical quantities.

- Dimension analysis from Maxwell's equations. Vector potential amplitude proportional to the frequency.

The relations (4.1.9) attribute a precise value of energy and momentum to a photon related to the electromagnetic wave only by the angular frequency, while no information is supplied about the photon vector potential which is basically indispensable in the classical electromagnetic description. As we have discussed in the previous Chapter 5 (5.1), in the electromagnetic theory, it is impossible to define the energy of an electromagnetic wave simply by its frequency while omitting the vector potential. Hence, if the photon is an integral particle with definite quantum energy then the exact quantization of its vector potential should be a natural consequence.

Obviously, the quantization procedure of the electromagnetic field, described briefly in Section (4.3.1), does not result effectively to the description of the vector potential of individual photons. Indeed, examining the link relations (4.3.14), obtained from (4.3.13) when $N = 1$, the amplitude of the vector potential for a single k mode photon with angular frequency ω_k is given by

$$\alpha_0 = \sqrt{\frac{\hbar}{2\varepsilon_0 \omega_k V}} \qquad (6.1.1)$$

Of course, for a large number of photons in a volume V, with short wavelengths compared to the dimensions of V, the summation transformation (3.5.44) resulting from the density of states helps to directly eliminate the external volume parameter V involved in the expression of the quantized vector potential amplitude. Consequently, one may argue that it's needless to enhance further the analysis on the definition of (6.1.1) to a single photon, since it is sufficient for the description of the physical situations studied in quantum electrodynamics (QED). Hence, by limiting the theoretical analysis of the equation (6.1.1), we miss a fundamental information on the photon vector potential that might ensure the link with the electromagnetic theory.

In fact, it is intriguing to deduce from (6.1.1), that the vector potential of a single photon depends on an independent free space variable V entailing that, for a given finite volume it is inversely proportional to the square root of the angular frequency. This is also puzzling from a physical point of view because, knowing that the current density $\vec{j}(\vec{r}, t)$ has the dimension of charge times m^{-2} s^{-1}, a dimension analysis of the general expression of the vector potential amplitude in equation (3.3.10) issued from Maxwell's equation shows that it should be proportional to the frequency, and consequently for a k-mode photon with angular frequency ω_k we should have

$$\alpha_0(\omega_k) \propto (\text{Constant} \times \omega_k) \rightarrow \xi\omega_k \qquad (6.1.2)$$

This is contradictory with the frequency dependence of (6.1.1), implying mathematically that in the case of a single photon state,

V cannot be a free external parameter in (6.1.1), but it has to be proportional to ω_k^{-3}

$$V_k \propto \left(\frac{1}{\omega_k^3} \right) \qquad (6.1.3)$$

The possible spatial expansion of a single photon state will be discussed later. Now we can have a close look to the photon vector potential amplitude quantization constant ξ.

• Wave equation for the photon vector potential

Based on (6.1.2), the fundamental physical quantities, energy, momentum, vector potential and wave vector, characterizing both the wave and particle nature of a single k mode photon state are all related to the angular frequency

$$\frac{E_k}{\hbar} = \frac{|\vec{p}_k|}{\hbar/c} = \frac{\alpha_0}{\xi} = |\vec{k}|c = \omega_k \qquad (6.1.4)$$

The expression of the photon vector potential may now be written in a general plane wave expression

$$\vec{\alpha}_{\omega_k}(\vec{r}, t) = \xi \omega_k [\hat{e} e^{i(\vec{k}\cdot\vec{r}-\omega_k t+\varphi)} + \hat{e}^* e^{-i(\vec{k}\cdot\vec{r}-\omega_k t+\varphi)}] \qquad (6.1.5)$$

whose propagation occurs within a period T, thus over a wavelength λ, and then repeated successively along the propagation axis.

Hence, the photon vector potential $\alpha_{\omega_k}(\vec{r}, t)$ has to satisfy the wave propagation equation

$$\vec{\nabla}^2 \vec{\alpha}_{\omega_k}(\vec{r}, t) - \frac{1}{c^2} \frac{\partial^2}{\partial t^2} \vec{\alpha}_{\omega_k}(\vec{r}, t) = 0 \qquad (6.1.6)$$

The second derivative with respect to time is proportional to the square of the amplitude

$$\frac{\partial^2}{\partial t^2} \vec{\alpha}_{\omega_k}(\vec{r}, t) = -\omega_k^2 \vec{\alpha}_{\omega_k}(\vec{r}, t) = -\left(\frac{\alpha_0}{\xi} \right)^2 \vec{\alpha}_{\omega_k}(\vec{r}, t) \qquad (6.1.7)$$

and the propagation equation can be written

$$[\alpha_0^2 + \xi^2 c^2 \vec{\nabla}^2] \vec{\alpha}_{\omega_k}(\vec{r}, t) = 0 \qquad (6.1.8)$$

entailing the interesting result that the photon vector potential amplitude can be expressed as an operator $\tilde{\alpha}_0$ proportional to ξ

$$\tilde{\alpha}_0 = -i\xi\, c\vec{\nabla} \qquad (6.1.9)$$

which is quite symmetrical with the relativistic Hamiltonian operator for a massless particle

$$\tilde{H} = -i\hbar c \vec{\nabla} \tag{6.1.10}$$

Let's apply $\tilde{\alpha}_0$ upon $\alpha_{\omega_k}(\vec{r}, t)$ along the propagation axis

$$\tilde{\alpha}_0 \vec{\alpha}_{\omega_k}(\vec{r}, t, \varphi) = i\xi ck\,\vec{\alpha}_{\omega_k}\left(\vec{r}, t, \varphi + \frac{\pi}{2}\right) \tag{6.1.11}$$

with φ the phase parameter.

But for the time variation, we also get

$$\frac{\partial}{\partial t}\vec{\alpha}_{\omega_k}(\vec{r}, t, \varphi) = \omega_k \vec{\alpha}_{\omega_k}\left(\vec{r}, t, \varphi + \frac{\pi}{2}\right) \tag{6.1.12}$$

The combination of the last two equations gives a linear time differential equation for the photon vector potential, valid within every wavelength interval successively along the propagation axis

$$i\xi \frac{\partial}{\partial t}\vec{\alpha}_{\omega_k}(\vec{r}, t) = \tilde{\alpha}_0\,\vec{\alpha}_{\omega_k}(\vec{r}, t) \tag{6.1.13}$$

The symmetry with the Schrödinger equation for a particle with a wave function $\phi(\vec{r}, t)$ is obvious

$$i\hbar \frac{\partial}{\partial t}\phi(\vec{r}, t) = H\,\phi(\vec{r}, t) \tag{6.1.14}$$

Furthermore, we also remark that when considering Heisenberg's energy time uncertainty principle

$$\delta E_k \delta t \geq \hbar \tag{6.1.15}$$

we deduce directly from the relations (6.1.4), a vector potential time uncertainty according to the relation

$$\delta \alpha_0 \delta t \geq \xi \tag{6.1.16}$$

It is worth noticing the symmetrical mathematical correspondence between the pairs $\{E, \hbar, \text{particle}\} \leftrightarrow \{\alpha_0, \xi, \text{wave}\}$ for a single photon.

- Photon wavelength dimensions following the experimental evidence

It is useful to recall that, in order to get the link equations (4.3.14), one has to consider the mean value of the left-hand side of (4.3.13), over a period T (or in space representation over a wavelength λ), while putting $N = 1$ to the right-hand side of (4.3.13). Notice also that the energy density of the electromagnetic field $\langle W_\omega \rangle_T$ and the energy flux obtained by Poynting vector $\langle \vec{S}_\omega \rangle_T$ in (3.4.8), are both calculated over a period T while they are completely independent on any external volume V.

It comes out that both these physical quantities, $\langle W_\omega \rangle_T$ and $\langle \vec{S}_\omega \rangle_T$, describing the energy density and flow of a "fraction" of the electromagnetic wave limited to its wavelength, depend only on the angular frequency ω, the amplitude of the vector potential, $\alpha_0(\omega)$, and on the vacuum electric permittivity ε_0.

Consequently, following the above aspects and considering the photon as an integral particle with a given energy, we may write using (3.2.29), (3.4.4) and (3.4.5) for a single k-mode photon

$$\int \varepsilon_0 \omega_k^2 \alpha_0^2(\omega_k)[\hat{\varepsilon} e^{i(\omega_k t - \vec{k} \cdot \vec{r} + \varphi)} + c.c]^2 d^3 r = \hbar \omega_k \qquad (6.1.17)$$

For the last relation to hold at any instant t, we deduce that the photon polarization unit vector should have at least two orthogonal components, $\hat{e}_1$ and $\hat{e}_2$, such as $\hat{\varepsilon} = \sigma_1 \hat{e}_1 + \sigma_2 \hat{e}_2$ with $|\sigma_1|^2 + |\sigma_2|^2 = 1$ and $\hat{e}_1 \cdot \hat{e}_2 = 0$.

According to (3.4.12), R and L hand circular polarization unit vectors, $\hat{e}_{L,R} = \frac{1}{\sqrt{2}}(\hat{e}_1 + i\hat{e}_2)$ are naturally appropriate to satisfy this condition and equation (6.1.17) becomes

$$\int 2\varepsilon_0 \omega_k^2 \alpha_0^2(\omega_k) \, d^3 r = \hbar \omega_k \qquad (6.1.18)$$

Indeed, it is generally accepted that the photons are characterized by L or R circular polarization.

The equation (6.1.18) is equivalent to the normalization of the energy of a classical electromagnetic plane wave over a wavelength, issued from Maxwell's equations, to Planck's expression of the quantized radiation energy. At that point, we recall the energy density equivalence between the classical and the quantum mechanical formulations given by (4.3.13), which for $N = 1$ is simply reduced to

$$2\varepsilon_0 \omega_k^2 \alpha_0^2(\omega_k) V_k = \hbar \omega_k \qquad (6.1.19)$$

where $\alpha_0(\omega_k)$ is the vector potential amplitude that may be attributed to a k mode photon and V_k, the corresponding quantization volume.

The comparison of the last two equations implies that V_k corresponds to an intrinsic property of the photon. From (6.1.2) and (6.1.19), we get

$$V_k = \left(\frac{\hbar}{2\varepsilon_0 \xi^2} \right) \omega_k^{-3} \qquad (6.1.20)$$

in agreement with the previous analysis resulting to (6.1.3).

Since it has been demonstrated that a photon cannot be considered as an integral entity within a region smaller than its wavelength λ, and since the directional character of photons has been demonstrated experimentally by Compton, the integration of (6.1.18) may be carried out in cylindrical coordinates (ρ, θ, z), where z is the propagation axis, along which the vector potential rotates perpendicularly. In a full period interval, $0 \leq z \leq \lambda$, the angular coordinate θ is related to z by $\theta = 2\pi z / \lambda$ and the integration may be reduced to the variables, z and ρ in the limits $0 \leq z \leq \lambda$ and $0 \leq \rho \leq \eta\lambda$, with η, a positive dimensionless constant characterizing the later spatial extension of the photon and permitting the following relation to hold

$$2\varepsilon_0 \omega_k^2 \alpha_0^2(\omega_k) \left(\frac{\eta^2}{2} \lambda^3 \right) = \varepsilon_0 \xi^2 \eta^2 \frac{(2\pi c)^3}{\omega_k^3} \omega_k^4 = \hbar \omega_k \qquad (6.1.21)$$

The product of the two constants η and ξ has the value

$$\eta\xi = \frac{1}{(2\pi)^{3/2}} \sqrt{\frac{\hbar}{\varepsilon_0 c^3}} \qquad (6.1.22)$$

In a first approximation, the last relation is almost independent of the integration coordinates of equation (6.1.18). For instance, integrating in spherical coordinates with $0 \leq r \leq \frac{\lambda}{2}$, one gets

$$2\varepsilon_0 \omega_k^2 (\xi \omega_k)^2 \int_0^{\lambda/2} 4\pi \chi^2 r^2 dr = \hbar \omega_k \qquad (6.1.23)$$

where χ^2 is equally a positive constant for the last equation to hold.

We obtain

$$\chi\xi = \frac{1}{(2\pi)^{3/2}} \sqrt{\frac{3\hbar}{\pi\varepsilon_0 c^3}} \tag{6.1.24}$$

The physical meaning of relations (6.1.22) and (6.1.24) is that Planck's constant is related to the vacuum's electric permittivity ε_0 by the intermediate of photon's intrinsic physical parameters.

• Wave-particle formalism

Following the expression (6.1.5), the photon vector potential can be written in a general wave-particle formalism

$$\vec{\alpha}_{\omega_{k\lambda}}(\vec{r}, t) = \omega_k[\xi\hat{\varepsilon}_{k\lambda}e^{i(\vec{k}\cdot\vec{r}-\omega_k t+\varphi)} + \xi^*\hat{\varepsilon}_{k\lambda}^* e^{-i(\vec{k}\cdot\vec{r}-\omega_k t+\varphi)}]$$

$$= \omega_k \vec{\Xi}_{k\lambda}(\omega_k, \vec{r}, t) \tag{6.1.25}$$

$$\tilde{\alpha}_{k\lambda} = \omega_k \lfloor\xi a_{k\lambda}\hat{\varepsilon}_{k\lambda}e^{i(\vec{k}\cdot\vec{r}-\omega_k t+\varphi)} + \xi^* a_{k\lambda}^+ \hat{\varepsilon}_{k\lambda}^* e^{-i(\vec{k}\cdot\vec{r}-\omega_k t+\varphi)}\rfloor$$

$$= \omega_k \tilde{\Xi}_{k\lambda}(a_{k\lambda}, a_{k\lambda}^+) \tag{6.1.26}$$

where for the quantum mechanical formalism, we have used the creation and annihilation operators $a_{k\lambda}^+$ and $a_{k\lambda}$ respectively for a k mode and λ polarization photon.

The general equation of the vector potential for the electromagnetic wave considered as a superposition plane waves writes

$$\vec{A}(\vec{r}, t) = \sum_{k,\lambda} \omega_k[\xi\hat{\varepsilon}_{k\lambda}e^{i(\vec{k}\cdot\vec{r}-\omega_k t+\varphi)} + \xi^*\hat{\varepsilon}_{k\lambda}^* e^{-i(\vec{k}\cdot\vec{r}-\omega_k t+\varphi)}]$$

$$= \sum_{k,\lambda} \omega_k \vec{\Xi}_{k\lambda}(\omega_k, \vec{r}, t) \tag{6.1.27}$$

and that of a large number of photons in the QED description is

$$\tilde{A} = \sum_{k,\lambda} \omega_k \lfloor\xi a_{k\lambda}\hat{\varepsilon}_{k\lambda}e^{i(\vec{k}\cdot\vec{r}-\omega_k t+\varphi)} + \xi^* a_{k\lambda}^+ \hat{\varepsilon}_{k\lambda}^* e^{-i(\vec{k}\cdot\vec{r}-\omega_k t+\varphi)}\rfloor$$

$$= \sum_{k,\lambda} \omega_k \tilde{\Xi}_{k\lambda}(a_{k\lambda}, a_{k\lambda}^+) \tag{6.1.28}$$

The vector potential with the quantized amplitude ξ in the wave expression (6.1.25) translates the individual wave property of the photon, while that of (6.1.26) its particle property, reflected through the vector potential amplitude operator. The mathematical representation of the wave and particle expression of the vector potential amplitude operator writes

$$\text{wave: } \{\tilde{\alpha}_0 = -i\xi c \vec{\nabla}; \ \tilde{\alpha}_0^* = i\xi^* c \vec{\nabla}\}$$

$$\text{particle: } \{\tilde{\alpha}_{0_{k\lambda}} = \xi \omega_k a_{k\lambda}; \ \tilde{\alpha}_{0_{k\lambda}}^* = \xi^* \omega_k a_{k\lambda}^+\} \quad (6.1.29)$$

We easily verify that when introducing $\tilde{\alpha}_{0_{k\lambda}}$ and $\tilde{\alpha}_{0_{k\lambda}}^*$ into the QED position and momentum operators expressions (4.3.11), one directly gets the corresponding position $\hat{Q}_{k\lambda}$ and momentum $\hat{P}_{k\lambda}$ operators

$$\hat{Q}_{k\lambda} = \sqrt{\varepsilon_0 V_k}(\tilde{\alpha}_{0_{k\lambda}} + \tilde{\alpha}_{0_{k\lambda}}^*) \quad \hat{P}_{k\lambda} = -i\omega_k \sqrt{\varepsilon_0 V_k}(\tilde{\alpha}_{0_{k\lambda}} - \tilde{\alpha}_{0_{k\lambda}}^*)$$
$$(6.1.30)$$

Using (6.1.20), the demonstration of Heisenberg's commutation relation is immediate

$$[\hat{Q}_{k\lambda}, \hat{P}_{k'\lambda'}] = -i\varepsilon_0 \omega_{k'}^2 \omega_k \sqrt{V_k V_{k'}}[(\xi a_{k\lambda} + \xi^* a_{k\lambda}^+), (\xi a_{k'\lambda'} - \xi^* a_{k'\lambda'}^+)]$$

$$= i\hbar \delta_{kk'} \delta_{\lambda\lambda'} \quad (6.1.31)$$

We can now deduce the particle properties of the photon by using the wave characteristics.

In fact, the particle properties, energy E_k and the momentum $\vec{P}_k$, of a k mode photon are not carried by a point particle and can be expressed in terms of the quantization volume V_k and the wave properties namely the dispersion relation, the vector potential and the electric and magnetic fields

$$E_k = 2\varepsilon_0 \omega_k^2 \alpha_{0_k}^2 V_k = \hbar \omega_k = \hbar kc \quad (6.1.32)$$

Considering equations (3.4.13) to (3.4.14) for circular polarization photons the momentum writes

$$\vec{P}_k = \int_{V_k} \varepsilon_0 \vec{E}^{(1)} \times \vec{B}^{(2)} d^3r$$

$$= \varepsilon_0 (\sqrt{2}\omega_k \alpha_{0_{k\lambda}}) \left(\frac{1}{c}\sqrt{2}\omega_k \alpha_{0_{k\lambda}}\right) V_k \frac{\vec{k}}{|\vec{k}|} = \hbar \vec{k} \quad (6.1.33)$$

The photon momentum has the dimensions of a mass m_k times the velocity, that is c in vacuum

$$|\vec{P_k}| = \hbar k = m_k c \qquad (6.1.34)$$

Combination of (6.1.32) and (6.1.34) gives

$$E_k = m_k c^2 \qquad (6.1.35)$$

showing that the energy–mass equivalence is a direct result of the wave-particle double nature of light which is extended to particles.

- Relation between the photon vector potential and the electron charge

Following the above analysis on the wave-particle nature of the photon, a quite interesting feature results when writing the energy equivalence (6.1.19), as follows

$$4\pi c \left(\frac{1}{4\pi c} 2\varepsilon_0 \omega_k^2 \alpha_0 (\omega_k) V_k \right) \alpha_0 (\omega_k) = 4\pi c\, Q\, \alpha_0 (\omega_k) = \hbar \omega_k$$

$$(6.1.36)$$

where the Q parameter has charge dimensions. Using (6.1.4) and (6.1.22), we obtain

$$Q^2 = \left(\frac{1}{4\pi c} 2\varepsilon_0 \omega_k^2 \alpha_0 (\omega_k) V_k \right)^2 = \frac{\eta^2}{4} (\hbar \varepsilon_0 c) \qquad (6.1.37)$$

Using for η, the experimental approximate value of $1/4$ discussed in Chapter 5 (5.5), it results immediately that $Q \sim 1.6\ 10^{-19}$ Coulomb, which is the electron charge, a physical constant, appearing naturally when considering the equivalence of the classical and quantum mechanical energy density of the electromagnetic field at a single photon level.

This entails that the physical origin of the electron charge is strongly related to the quantum of the electromagnetic field over a period, the photon.

Based on this result, we can go further in the definitions of both η and ξ constants by the intermediate of the fine structure constant α_{FS}, as defined in the equation (4.6.4), by equating the square of the electron charge to (6.1.37)

$$\frac{\eta^2}{4} (\hbar \varepsilon_0 c) \approx 2\alpha_{\mathrm{FS}} (\hbar \varepsilon_0 c) \qquad (6.1.38)$$

getting

$$\eta \approx \sqrt{8\alpha_{\mathrm{FS}}} \qquad (6.1.39)$$

Consequently, we can have an approximate idea of the numerical value of the vector potential amplitude quantization constant from (6.1.36) or the coupling of (6.1.22) and (6.1.39)

$$4\pi\, c\, Q\alpha_0(\omega_k) \approx 4\pi\, c\, e\xi\omega_k \approx \hbar\omega_k \qquad (6.1.40)$$

$$\xi \propto \frac{1}{(2\pi)^{3/2}} \sqrt{\frac{\hbar}{8\alpha_{\mathrm{FS}}\varepsilon_0 c^3}} = \frac{\hbar}{4\pi e c} = 1.747\ 10^{-25}\ \mathrm{Volt\ m^{-1}s^2}$$

$$(6.1.41)$$

Appropriate experiments could attribute more precise values to η and ξ.

6.2. Quantum Vacuum Representation

Let us come back to the two values of the Hamiltonian, (5.2.1) and (5.2.3) which result simply from the commutation of $A^*_{k\lambda}$ and $A_{k\lambda}$ in equation (4.3.9), and let us deal with the fundamental problem of the mathematical origin of the $\hbar\omega/2$ term.

As it is well known, a simple commutation between the vector potential amplitude and its complex value, which has absolutely no influence in the classical expression of the electromagnetic field energy (4.3.9), proves to be of crucial importance for the quantum mechanical Hamiltonian. We can now analyze the reason for which the radiation Hamiltonian (4.3.18) cannot be obtained when the link equations (4.3.14) are introduced in the expression of the electromagnetic field energy (4.3.9).

It is important noticing that, for a harmonic oscillator of a particle of mass m and momentum $\vec{p} = m\, d\vec{q}/dt$, the transition from the classical expression of energy (4.2.20) to the Hamiltonian (4.2.25)

$$E_{\mathrm{HO}} = \frac{1}{2}(P^2 + \omega^2 Q^2) \rightarrow \tilde{H}_{\mathrm{HO}} = \frac{1}{2}(\hat{P}^2 + \omega^2 \hat{Q}^2) = \hbar\omega\left(a^+ a + \frac{1}{2}\right)$$

$$(6.2.1)$$

where $\hat{P} = i\sqrt{\frac{\hbar\omega}{2}}(a^+ - a)$ and $\hat{Q} = \sqrt{\frac{\hbar}{2\omega}}(a^+ + a)$, is immediate and needs no commutation operations between the canonical variables P and Q.

Consequently, the harmonic oscillator Hamiltonian $\tilde{H}_{\text{HO}}$ is a direct result expressing a perfect correspondence between the classical canonical variables of momentum and position, P and Q respectively, and the quantum mechanical Hermitian operators $\hat{P}$ and $\hat{Q}$.

Conversely, this is not the case for the electromagnetic field because commutations between $Q_{k\lambda}$ and $P_{k\lambda}$ occur during the transition from the expression (4.3.9) to (4.3.12). Therein, we recall that Heisenberg's commutation relation $[\hat{Q}_{k\lambda}, \hat{P}_{k'\lambda'}] = i\hbar\delta_{kk'}\delta_{\lambda\lambda'}$ is a fundamental concept of quantum mechanics which should not be ignored, when transiting from classical variables to quantum mechanical operators. This is exactly the reason for which the equation (4.3.12) cannot be considered as the electromagnetic field energy expression for establishing the quantized radiation Hamiltonian since it is not equivalent, from the quantum mechanical point of view, to equation (4.3.9).

In fact, introducing the vector potential amplitude expressions of (4.3.11) in the equation (4.3.9)

$$\left\{ A_{k\lambda} = \frac{(\omega_k Q_{k\lambda} + iP_{k\lambda})}{2\omega_k\sqrt{\varepsilon_0 V}}; \ A_{k\lambda}^* = \frac{(\omega_k Q_{k\lambda} - iP_{k\lambda})}{2\omega_k\sqrt{\varepsilon_0 V}} \right\}$$

$$\rightarrow E = 2\varepsilon_0 V \sum_{k,\lambda} \omega_k^2 |A_{k\lambda}|^2 \qquad (6.2.2)$$

and keeping the commutation terms appearing in the calculation one gets, instead of the expression (4.3.18), i.e., $E_{\text{EM}} = \frac{1}{2}\sum_{k,\lambda}(P_{k\lambda}^2 + \omega_k^2 Q_{k\lambda}^2)$, the complete expression for the energy of for the electromagnetic field

$$E_{\text{EM}} = \frac{1}{2}\sum_{k,\lambda}(P_{k\lambda}^2 + \omega_k^2 Q_{k\lambda}^2) \pm i\omega_k[Q_{k\lambda}, P_{k\lambda}] \qquad (6.2.3)$$

where the (+) sign is obtained when the starting equation (4.3.9) is considered to be in the "normal ordering", $2\varepsilon_0 V \sum_{k,\lambda} \omega_k^2 A_{k\lambda}^* A_{k\lambda}$, and the (−) one when $2\varepsilon_0 V \sum_{k,\lambda} \omega_k^2 A_{k\lambda} A_{k\lambda}^*$.

Equation (6.2.3) is a rigorous mathematical result leading to the radiation Hamiltonians

$$\tilde{H}_{\text{EM}} = \frac{1}{2} \sum_{k,\lambda} (\hat{P}_{k\lambda}^2 + \omega_k^2 \hat{Q}_{k\lambda}^2) \pm i\omega_k [\hat{Q}_{k\lambda}, \hat{P}_{k'\lambda'}]$$

$$= \left(\begin{array}{cc} \displaystyle\sum_{k,\lambda} \hat{N}_{k\lambda} \, \hbar\omega_k & (+) \\ \displaystyle\sum_{k,\lambda} (\hat{N}_{k\lambda} + 1)\hbar\omega_k & (-) \end{array} \right) \qquad (6.2.4)$$

where the upper summation of the last part corresponds to the $(+)$ sign, hence to the normal ordering of the vector potential amplitude and its conjugate, while the lower summation to the $(-)$ sign. This is the correct mathematical way to get identical results with the Hamiltonian (6.2.4) when the link equations (4.3.14) are introduced in the classical energy expression (4.3.9).

Consequently, one may wonder whether the vacuum contribution to the quantized electromagnetic field Hamiltonian, i.e., the term $\sum_{k,\lambda} \frac{1}{2}\hbar\omega_k$ of equation (4.3.18) obtained when neglecting arbitrarily the commutation $\pm i\omega_k [Q_{k\lambda}, P_{k\lambda}]$ without respecting Heisenberg's relation, can really represent a physical state.

From a pure mathematical point of view the classical electromagnetic field energy may be equivalent to that of an ensemble of quantized harmonic oscillators when considering the mean value of the "normal ordering" and "anti-normal ordering" expressions getting

$$E_{EM} = \frac{1}{2} \left[2\varepsilon_0 V \sum_{k,\lambda} \omega_k^2 A_{k\lambda}^* A_{k\lambda} + 2\varepsilon_0 V \sum_{k,\lambda} \omega_k^2 A_{k\lambda} A_{k\lambda}^* \right]$$

$$= \frac{1}{2} \left[\frac{1}{2} \sum_{k,\lambda} (P_{k\lambda}^2 + \omega_k^2 Q_{k\lambda}^2) + i\omega_k [Q_{k\lambda}, P_{k\lambda}] \right.$$

$$\left. + \frac{1}{2} \sum_{k,\lambda} (P_{k\lambda}^2 + \omega_k^2 Q_{k\lambda}^2) - i\omega_k [Q_{k\lambda}, P_{k\lambda}] \right]$$

$$= \frac{1}{2} \sum_{k,\lambda} (P_{k\lambda}^2 + \omega_k^2 Q_{k\lambda}^2) \rightarrow \frac{1}{2} \sum_{k,\lambda} (\hat{P}_{k\lambda}^2 + \omega_k^2 \hat{Q}_{k\lambda}^2)$$

$$= \sum_{k,\lambda} (\hat{N}_{k\lambda} + \frac{1}{2}) \hbar \omega_k \qquad (6.2.5)$$

However, the physical reasons for considering the above mean value are rather elusive. Nevertheless, as it has been analyzed in Section (5.4), the zero-point energy of the electromagnetic field Hamiltonian as a constant value, either $\sum_{k,\lambda} \frac{1}{2} \hbar \omega_k$ or $\sum_{k,\lambda} \hbar \omega_k$, is not really involved directly in the electron-vacuum interactions. Therein, it has been pointed out by many authors that the mathematical and physical basis for the description of the electron-vacuum interactions issues from the properties of the photon creation and annihilation operators, $a_{k\lambda}^+$ and $a_{k\lambda}$ respectively.

Consequently, if the Hamiltonian $\sum_{k,\lambda} \hat{N}_{k\lambda} \hbar \omega_k$ obtained when using the normal ordering (+) in the radiation Hamiltonian (6.2.4) represents the physical state of a photon ensemble then what could be the expression for the quantum vacuum? We can now investigate what could be the vacuum expression in this case. According to the relations (6.1.4), for $\omega_k = 0$, the vector potential, the energy, and the momentum of a k mode photon vanish. However, in complete absence of energy, the field $\Xi_{k\lambda}$ in (6.1.25) to (6.1.28) does not vanish and reduces to the vacuum field that can also be described as a quantum mechanical operator

$$\vec{\Xi}_{0_{k\lambda}} = \xi \hat{\varepsilon}_{k\lambda} e^{i\varphi} + \xi^* \hat{\varepsilon}_{k\lambda}^* e^{-i\varphi} \qquad (6.2.6)$$

$$\tilde{\Xi}_{0_{k\lambda}} = \xi a_{k\lambda} \hat{\varepsilon}_{k\lambda} e^{i\varphi} + \xi^* a_{k\lambda}^+ \hat{\varepsilon}_{k\lambda}^* e^{-i\varphi} \qquad (6.2.7)$$

Thus, $\tilde{\Xi}_{0_{k\lambda}}$ is a real entity of the vacuum state having the dimensions Volt m^{-1} s^2 corresponding to a vector potential per angular frequency and implying an electric nature of the quantum vacuum, as in the case of the classical electromagnetic description. In fact, as we have seen in (3.2.13) and (3.2.15), since the very beginning of the electromagnetic theory, the vacuum state has been characterized by an impendence $R_{\text{vacuum}} = \sqrt{\frac{\mu_0}{\varepsilon_0}} \approx 120\pi$ Ω, a magnetic permeability $\mu_0 = 4\pi \times 10^{-7} H\,m^{-1}$ and an electric permittivity $\varepsilon_0 = 8.85 \times 10^{-12} F\,m^{-1}$.

The vacuum field $\tilde{\Xi}_{0k\lambda}$ constitutes the "skeleton" of photons and consequently the energy, and by the same token, the mass appear to be the direct result of real vacuum vibrations. Under specific physical conditions, photons and elementary particles might emerge from the quantum vacuum.

Obviously, $\tilde{\Xi}_{0k\lambda}$ is a dynamic entity capable of inducing electronic transitions in matter since it is described by a function of $a_{k\lambda}^{+}$ and $a_{k\lambda}$ operators according to (6.2.7).

Furthermore, following equation (6.1.41), the electric character of the vacuum expressed through the constant ξ makes that every charge moving in free space with an acceleration γ will experience an electric potential U_{vacuum} due to the vacuum electric nature

$$U_{\text{vacuum}} = \gamma\xi \qquad (6.2.8)$$

Finally, introducing (6.1.20) in (5.3.5), we obtain the expectation value of the electric field in vacuum

$$\langle E(\vec{r})^2 \rangle_{\text{vacuum}} = \sum_{k,\lambda} \omega_k^4 \xi^2 \to 0 \quad for \ \omega_k(\text{vacuum}) \to 0 \qquad (6.2.9)$$

Hence, in this description the QED singularities, related to the infinite vacuum energy and the infinite expectation value of the photons electric field in the vacuum state are lifted.

6.3. The Quantum Vacuum Field Effects

Taking into account the wave-particle formalism developed in the previous chapter, we examine here how the effects due to the quantum vacuum, described briefly in Section 4 and whose difficulties are encountered in Section 5, can now be described in a more coherent and comprehensive mathematical approach.

• Spontaneous emission

For the description of the electron–vacuum interaction and in order to overcome the difficulty arising from the (5.4.4), a "vacuum action" operator corresponding to an interaction Hamiltonian per angular frequency between the vacuum field $\tilde{\Xi}_{0_{k\lambda}}$ and an atomic electron of mass m_e and charge e can now be defined like the third term of the Hamiltonian (4.4.3):

$$H_{\omega_k} = -i\hbar \frac{e}{m_e} \tilde{\Xi}_{0_{k\lambda}} \cdot \vec{\nabla} \qquad (6.3.1)$$

Thus, the amplitude of the transition probability per angular frequency between an initial state $|\Psi_i, 0\rangle$ with total energy $E_i = \hbar\omega_i$, corresponding to an atom in the state $|\Psi_i\rangle$ with energy E_i in vacuum $|\ldots, 0_{k\lambda}, \ldots, 0_{k'\lambda'}, \ldots\rangle$, and a final state $|\Psi_f, n_{k\lambda}\rangle$ with total energy $E_{f,n_{k\lambda}} = \hbar\omega_f + n_{k\lambda}\hbar\omega_k$ representing the atom in the state $|\Psi_f\rangle$ with energy $E_f = \hbar\omega_f$ in the presence of $n_{k\lambda}$ photons, can be expressed in first order time dependent perturbation theory using (4.5.5)

$$c_{\omega_k}^{(1)}(t) = -\frac{e}{m_e} \int_0^t [\langle \Psi_f, n_{k\lambda} | \xi \vec{\varepsilon}_{k\lambda} \cdot \vec{\nabla} a_{k\lambda} | \Psi_i, 0\rangle e^{i(\omega_i - \omega_f - n_{k\lambda}\omega_k)t'}$$
$$+ \langle \Psi_f, n_{k\lambda} | \xi^* \vec{\varepsilon}_{k\lambda}^* \cdot \vec{\nabla} a_{k\lambda}^+ | \Psi_i, 0\rangle e^{i(\omega_i - \omega_f - n_{k\lambda}\omega_k)t'}] dt' \qquad (6.3.2)$$

Since $a_{k\lambda}|\Psi_i, 0\rangle = 0$, using the fundamental definition of the momentum operator (4.1.12), as well as Heisenberg's equation of motion (4.5.22b), the scalar product of equation (6.3.2) corresponding to the creation operator $a_{k\lambda}^+$ writes

$$\langle \Psi_f, n_{k\lambda} | \xi^* \vec{\varepsilon}_{k\lambda}^* \cdot \vec{\nabla} a_{k\lambda}^+ | \Psi_i, 0\rangle = -\xi^* \delta_{1,n_{k\lambda}} m_e \omega_{if} \vec{\varepsilon}_{k\lambda}^* \cdot \vec{r}_{if}/\hbar \qquad (6.3.3)$$

with $\vec{r}_{if} = \langle \Psi_f | \vec{r} | \Psi_i \rangle$, $\omega_{if} = \omega_i - \omega_f$ and considering the expression of $\xi \propto \frac{\hbar}{4\pi ec}$ from (6.1.41), one gets with an equivalent calculation as in (4.5.24), the spontaneous emission rate in the elementary solid angle $d\Omega$.

$$W_{if} = \frac{1}{3\hbar c^3} \frac{e^2}{4\pi^2 \varepsilon_0} \omega_{if}^3 |\vec{r}_{if}|^2 d\Omega \qquad (6.3.4)$$

The calculation shows that the spontaneous emission is mainly due to the creation operator $a_{k\lambda}^+$, which here is involved in the quantum vacuum expression $\tilde{\Xi}_{0_{k\lambda}}$ permitting to establish an interaction Hamiltonian.

• Lamb shift

In Section 4.6., we have calculated the energy level displacement
due to the vacuum interaction by employing a "particular calculation" consisting of neglecting infinite quantities and imposing arbitrary integration limits. Let us now have a much simpler approach
on one of the ways the vacuum state could interact with the atomic
energy levels.

In a first approximation, the energy level displacements of
the electron bounded states can be seen as a topological effect of the
vacuum radiation pressure upon the electronic orbitals. In fact, the
motion of a bounded state electron with charge e, whose energy is
E_{nlj}, in the vacuum field $\tilde{\Xi}_{0_{k\lambda}}$ can be characterized by Bohr's angular
frequency

$$\omega_{\text{nlj}} = \frac{E_{\text{nlj}}}{\hbar} = \frac{R_{\infty}}{n^2\hbar} \qquad (6.3.5)$$

Where n, l, j are the quantum numbers of the corresponding
orbital entailing the rise in the electron frame of a vector potential
amplitude

$$\alpha_{0(\text{nlj})} = \xi\omega_{(\text{nlj})} \qquad (6.3.6)$$

corresponding to "vacuum photons" with energy

$$4\pi ec\xi\omega_{\text{nlj}} = \hbar\omega_{\text{nlj}} \qquad (6.3.7)$$

Hence, due to its periodic motion an electron experiences in its
frame vacuum photons whose radiation pressure per unit surface
writes

$$dP(\text{vac}) = \sum_{\lambda} \varepsilon_0 |E_0^{(R,L)}|^2 d\Omega \qquad (6.3.8)$$

Where $|E_0^{(R,L)}|$ is the electric mean field intensity of the L and R
hand polarized vacuum photons seen by the electron which can be
expressed by the vector potential mean components, using (3.4.13),

$$|E_0^{(R)}|^2 = \omega_{\text{nlj}}^2(|A_{x_0}^{(R)}|^2 + |A_{y_0}^{(R)}|^2) = \omega_{\text{nlj}}^2(|A_{x_0}^{(L)}|^2 + |A_{y_0}^{(L)}|^2)$$

$$= \omega_{\text{nlj}}^2(2\xi^2\omega_{\text{nlj}}^2) \qquad (6.3.9)$$

Consequently, using (6.3.9) and (6.1.41), the summation of (6.3.8)
over the two circular polarizations, R and L, and over the whole solid

angle gives the total vacuum radiation pressure

$$P(\text{vac}) = 4\pi\varepsilon_0(4\xi^2\omega_{\text{nlj}}^4) \approx \frac{\hbar\omega_{\text{nlj}}^4}{4\alpha\pi^2 c^3} \qquad (6.3.10)$$

Hence, if the electron subsists in the effective volume $V_{\text{nlj}}(\text{eff})$ of the nlj orbital, the corresponding energy shift due to vacuum radiation pressure writes

$$\delta E = P(\text{vac})\,V_{\text{nlj}}(\text{eff}) \approx \frac{\hbar\omega_{\text{nlj}}^4}{4\alpha\pi^2 c^3}V_{\text{nlj}}(\text{eff}) \qquad (6.3.11)$$

A delicate operation consists of defining $V_{(\text{nlj})}(\text{eff})$. The nS electronic orbitals for example are characterized by a density probability distribution decreasing smoothly with an exponential expression. Furthermore, for the excited electronic nS states, the radial wave functions contain negative values and the electron probability density distribution involves a significant space area with zero values. Hence, in the case of atomic hydrogen for the spherically symmetrical electronic orbitals nS, in order not to take into account the space regions where the probability density is zero, the effective volume may be written in a first approximation

$$V_{(\text{nS})}(\text{eff}) \approx \frac{1}{n}\frac{4}{3}\pi|\langle\psi_{\text{ns}}|r|\psi_{\text{ns}}\rangle|^3 \qquad (6.3.12)$$

and following the last two equations, the corresponding frequency of the energy shift

$$\nu_{\text{nS}} = \frac{\delta E_{(\text{nS})}}{h} \approx \frac{\omega_{\text{nS}}^4}{6\alpha\pi^2 c^3}\frac{1}{n}|\langle\psi_{\text{ns}}|r|\psi_{\text{ns}}\rangle|^3 \qquad (6.3.13)$$

Putting $\omega_{\text{nS}} = \frac{R_\infty}{n^2\hbar}$ with $R_\infty = 13.606\,\text{eV}$, and considering the cube of the mean value of the distance in the hydrogen electronic orbitals nS

$$|\langle r\rangle|_{n,l=0}^3 = \left|\int \Psi_{(n,l=0)}r\Psi_{(n,l=0)}^* d^3 r\right|^3 = a_0^3\left(\frac{3n^2}{2}\right)^3 \qquad (6.3.14)$$

with $a_0 = 0.53\ 10^{-10}$ m, being the first Bohr radius, the frequency of the energy shift writes

$$\nu_{\text{nS}} \approx \frac{1}{n^3}\frac{9\,R_\infty^4 a_0^3}{16\,\alpha\pi^2 c^3\hbar^4} \qquad (6.3.15)$$

Significant Lamb shifts have been observed for the nS orbitals (n being the principal quantum number) having a spherical density probability distribution and zero orbital momentum $l = 0$.

We obtain the Lamb shifts frequencies for the hydrogen nS levels:

1S: $\sim$7.96 GHz (8.2), 2S: $\sim$1000 MHz (1040), 3S: $\sim$296 MHz (303),
4S: $\sim$124 MHz,

where the experimental values are given in parenthesis. Consequently, the energy shifts of the nS levels of the hydrogen atom interacting with vacuum can be estimated with a rather good approximation for such a simple calculation, within a physically comprehensive approach.

• Cosmic vacuum energy density

In Chapter 5 (5.3), we have pointed out that recent astronomical observations revealed the vacuum energy density in the universe to be many orders of magnitude lower than that predicted by QED even when considering cut-offs for the high frequencies near the visible spectrum for the integration of equation (5.3.3).

In fact, the studies on the dark energy and the arising constraints have stimulated the creation of new models to remedy the theoretical cosmology difficulties. Carroll *et al.* have shown that the modelling of the dark energy as a fluid has led to the classical phantom description which revealed to be instable from both quantum mechanical and gravitational point of view. Some proposals based on phantom cosmology have introduced more consistent dark energy models such as Fanaoni's quantum effects in general relativity, Nojiri's nonlinear gravity-matter interaction, Abdalla's revised gravity and Elizalde's effective phantom phase and holography among many others.

On the other hand, various studies on the contribution of the quantum vacuum state to the dark energy have been carried out based on both quantum chromodynamics (QCD) and QED descriptions like those of Frieman, Silvestri, and more recently Labun's works. However it has been pointed out by many authors that our understanding of the QCD vacuum state still remains very elusive. Conversely, QED vacuum offers a more comprehensive framework and it is worthy to analyze the eventual contribution of the electromagnetic field lower energy level to the dark energy, starting from the basic principles of the classical electromagnetism and QED theories.

According to the relation (6.1.20), for photons with very low angular frequencies, the corresponding spatial extension gets extremely big. For $\omega_k \to 0$, the wavelength of a k mode photon tends to cosmic dimensions while the energy, the vector potential and the momentum tend to extremely low values. However, even for $\omega_k = 0$, the field $\vec{\Xi}_{k\lambda}(\omega_k, \vec{r}, t)$ does not vanish and reduces to $\vec{\Xi}_{0k\lambda}$ expressed by (6.2.6) and (6.2.7), filling the whole space. Although in the case of an infinite universe, this state corresponds to zero energy, $\vec{\Xi}_{0k\lambda}$ is a real entity having the dimensions Volt m^{-1} s^2. We have thus pointed out that $\vec{\Xi}_{0k\lambda}$ can be identified to a quantum vacuum field with an electric nature. A k mode photon may be considered as a vacuum soliton since it is composed of $\vec{\Xi}_{0k\lambda}$, rotating with an angular frequency ω_k.

For a finite, universe behaving as a cosmic cavity, ω_k in (6.1.20) gets infinitely small but not zero. Hence, we may assume the universe to be filled by electromagnetic waves with cosmic dimensions. Under these conditions, with the recent astrophysical data on the possible dimensions of the universe, an extremely high number of cosmic photons of the background vacuum energy, roughly of the order of 10^{100} could contribute to the energy density whose measured value is about 10^{-9} J/m^3.

Finally, in particular situations the various distributions of mass and charges in the universe may induce fluctuations of $\vec{\Xi}_{0k\lambda}$, giving rise to higher frequency electromagnetic waves that may lay in the range of extremely long wavelength radio waves up to the microwaves, that might contribute to the microwaves irregularities observed in space.

6.4. Conclusion Remarks

In this chapter we have advanced theoretical elaborations in order to ensure a coherent link between the electromagnetic wave theory and Quantum Electrodynamics with the purpose to establish a non-local wave-particle description of the photon as required by the experiments. We have started from the fact that the vector potential amplitude of a single photon state is not defined in QED (6.1.1) since it depends on an arbitrary external volume parameter V. Taking into account that according to Maxwell's theory the vector potential amplitude should be proportional to the frequency we elaborated a non-local wave-particle formalism. The quantization constant of the vector potential amplitude is strongly related to the quantum vacuum. Under these conditions, Heisenberg's commutation relations as well as the electron charge, a physical constant, appear naturally from the non-local expression of the photon energy. Hence, photons (and perhaps electrons and other particles) appear to be "corpuscules" (solitons) issued from the electromagnetic quantum vacuum. Finally, some quantum vacuum field effects such as the spontaneous emission, Lamb's shift and the cosmological vacuum energy density have been discussed showing the consistency of the elaborations. Consequently, advanced experiments are indispensable for the determination of the photon vector potential amplitude constant, which also characterizes the quantum vacuum.

Bibliography

1. M.C.B. Abdalla, S. Nojiri and S.D. Odintsov, Consistent modified gravity: dark energy, acceleration and the absence of cosmic doomsday, *Class. Quantum Grav.* **22**(5) (2005) L35–L42.

2. A.I. Akhiezer and B.V. Berestetskii, *Quantum electrodynamics*, New York: Interscience Publishers. 1965.

3. C.L. Andrews, *Optics of the electromagnetic spectrum*, Englewood cliffs, New Jersey: Prentice-Hall Inc. 1960.

4. A. Aspect, *The wave-particle dualism,* S. Diner, D. Fargue, G. Lochak, & F. Selleri (Eds.) Dordrecht: DReidel Publication Company. 1984, pp. 297–330.

5. G. Auletta, *Foundations and interpretation of quantum mechanics,* Singapore: World Scientific. 2001.

6. H.A. Bethe, The electromagnetic shift of energy levels, *Phys. Rev.* **72**(4) (1947), 339–341.

7. S.M. Carroll, M. Hoffman and M. Trodden, Can the dark energy equation-of-state parameter w be less than -1? *Phys. Rev.* D**68** (2003), DOI: org/10.1103/PhysRevD.68.023509.

8. L. de Broglie, *Une tentative d'interprétation causale et non linéaire de la mécanique ondulatoire: la théorie de la double solution,* Paris: Gauthier-Villars. 1956.

9. C.P. Enz, Is the zero-point energy real? In *Physical reality and mathematical description*, C.P. Enz, & J. Mehra (Eds.) Dordrecht: DReidel Publication Company. 1974.

10. V. Fanaoni, *Nucl. Phys. B* **703** (2004) 454.

11. R. Feynman, *The strange theory of light and matter*, Princeton, New Jersey: Princeton University Press. 1988.
12. J. Frieman, M. Turner and D. Huterer, Dark energy and the accelerating universe, *Ann. Rev. Astron. Astrophys.* **46** (2008) 385–432.
13. R.K. Hadlock, Diffraction patterns at the plane of a slit in a reflecting screen, *J. Appl. Phys.* **29**(6) (1958) 918–920.
14. T. Hey, *The new quantum universe*, Cambridge: Cambridge University Press. 2003.
15. L. Labun and J. Rafelski, Vacuum structure and dark energy, *Int. J. Mod. Phys. D* **19**(14) (2010) 2299.
16. C. Meis, Photon wave-particle duality and virtual electromagnetic waves, *Found. Phys.* **27**(6) (1997) 865–873.
17. C. Meis, Zero-point radiation field revisited, *Phys. Essays* **12**(1) (1999).
18. C. Meis, Electric potential of the quantum vacuum, *Phys. Essays* **22**(1) (2009).
19. C. Meis, Vector potential quantization and the quantum vaccum, *Physics Research International* **187432** (2014).
20. P.W. Milonni, *The quantum vacuum*, London: Academic Press Inc. 1994.
21. S. Nojiri and S.D. Odintsov, Gravity assisted dark energy dominance and cosmic acceleration, *Phys. Lett. B* **599** (2004) 137.
22. T. Padmanabhan, Cosmological constant — the weight of the vacuum, *Phys. Rep.* **380**(5–6) (2003) 235–320.
23. Yu.P. Rybakov and B. Saha, Soliton model of atom, *Found. Phys.* **25**(12) (1995) 1723–1731.
24. A. Silvestri and M. Trodden, Approaches to understanding cosmic acceleration, *Rep. Prog. Phys.* **72**(9) (2009), DOI: org/10.1088/0034-4885/72/9/096901.
25. L.J. Wang, X.Y. Zou and L. Mandel, Experimental test of the de Broglie guided-wave theory for photons, *Phys. Rev. Lett.* **66** (1991) 1111.
26. S. Weinberg, The cosmological constant problem, *Rev. Mod. Phys.* **61** (1989) 1.
27. X.Y. Zou, L.J. Wang, T. Grayson and L Mandel, Can an "empty" de Broglie pilot wave induce coherence? *Phys. Rev. Lett.* **68**(25) (1992) 3667.
28. C.M. Wilson *et al.*, Observation of the dynamical Casimir effect in a superconducting circuit, *Nature* **479** (2011) 376–379.

Chapter 7

Epilogue

The mathematical expressions of a valid theory should correctly describe the physical mechanisms involved in a complete ensemble of experimentally observed phenomena. When many theories, each based on different concepts, interpret only partially some measurable observables, then it becomes obvious that they cannot be epistemologically valid, entailing that they have to be improved, if possible, or abandoned. Hence, the historical evolution over twenty five centuries of the concepts on light's nature followed the sequence: corpuscles, ray optics (without having any idea on the nature of the particles composing the rays), wave optics (with no idea about the nature of the waves), electromagnetic waves theory and finally quantum particle theory. Despite all these evolutions, the general picture of the nature of light is not satisfactory and Bohr's *complementarity principle* was, and still remains, the artificial mask behind which is hidden the difficulty of our comprehension of the simultaneous wave-particle concept. However, the *mutual exclusiveness* of Bohr's *complementarity principle*, according to which light has both wave and particle natures but only one of those is expressed explicitly in a given physical situation, is strongly put in doubt by recent experiments.

In this book, we have tried to resume selected chapters of both classical electromagnetism and quantum electrodynamics (QED), in order to help the reader have an opinion on how the nature of light is represented through each theory. It is of outstanding importance to note that in both representations, the light and the vacuum are strongly related entities.

Classical electrodynamics issued from Maxwell's equations revealed the necessity of introducing the notion of *volume* for an electromagnetic wave to stand, entailing precise values of cut-off wavelengths to account for the shape and dimensions of the surrounding space. Conversely, QED theory considers light to be composed of *point particles*, disregarding the conceptual question on how the frequency of oscillating electric and magnetic fields over a wavelength may be attributed to a point particle. The theoretical basis for establishing the QED mathematical expressions for a *point particle* is the classical and the quantum mechanical energy density equivalence (4.3.13), in which the *mean value of the energy is taken over a wavelength and attributed to a point particle*. Obviously, the incapacity of QED to advance a comprehensive description of a single wave-particle photon state relies on the point particle concept.

In Chapter 5 we have analysed the QED main theoretical difficulties and singularities. The zero-point energy of the electromagnetic field, as deduced in the Hamiltonian (4.3.18) issues from the mathematical artifact related to the dropping of the permutation of the canonical variables of position and momentum during the quantization procedure, without respecting Heisenberg's relations and results to a quantum vacuum having infinite energy, an unobserved experimental singularity. Furthermore, the measurable effects attributed to the quantum vacuum, i.e., the spontaneous emission, the atomic energy shifts and the Casimir effect may quite well be interpreted using aspects of electrodynamics and without referring at all to the zero-point energy of the Hamiltonian (4.3.18). Finally, the contribution of the electromagnetic field vacuum energy to the cosmological constant, as calculated in QED, is in contradiction with the astronomical observations by many orders of magnitude. All these strongly entail that the zero-point energy of the Hamiltonian (4.3.18) may not correspond to a real physical state of the electromagnetic field.

On the experimental front we have seen that the photoelectric effect, traditionally presented as a demonstration of the particle nature of light is pretty well interpreted, by applying classical electromagnetism using the wave concept. On the other hand, Young's diffraction experiments, initially supporting the wave theory, were quite well interpreted using particle principles. Finally, we presented some particular experiments, requiring for their interpretation the notion of the simultaneous wave particle of light, and which can be hardly interpreted either by classical electromagnetism alone or the QED theory.

The natural way out of the wave-particle dilemma and the QED shortcomings is to consider that *the photon is not a point particle* but rather a segment (soliton) of the electromagnetic field, hence with a real wave function, composed of a quantized vector potential with definite amplitude oscillating over a wavelength. This is the essential concept of the theoretical elaborations we have advanced in Chapter 6, starting from the electromagnetic field vector potential amplitude issued from Maxwell's equations, which is proportional to the angular frequency. Consequently, for the energy of the photon, considered as an integral and indivisible entity, to correspond to a definite value $\hbar\omega$, the vector potential amplitude has to be quantized to a single photon level $\alpha_0 = \xi\omega$. Next, the equivalence between the classical and the quantum mechanical electromagnetic field energy density (4.3.13) results inevitably to a photon wavelength spatial expansion $V \propto \omega^{-3}$ according to the relation (6.1.20). The equations (6.1.4) relate the particle (energy, momentum) and wave (frequency, vector potential amplitude, wave vector) physical quantities characterizing the wave-particle nature of light. In the extension of these calculations, the vacuum appears naturally to have an electric nature, composed of a universal field $\tilde{\Xi}_{0_{k\lambda}}$ with the amplitude ξ, which is the essence of the photon vector potential amplitude. Under these conditions, the photon can be seen as a quantum vacuum soliton. The mass–energy equivalence is also a direct result of the wave particle nature of the photon. Furthermore, as shown in (6.1.36), the electron charge, a physical constant, appears naturally when considering the equivalence between the energy of a segment of the electromagnetic field restricted to the wavelength and the quantum expression $\hbar\omega$. From (6.1.41), the electron charge may be expressed only through the vacuum parameters $\{\varepsilon_0, \mu_0, \xi\}$ and Planck's constant: $e \propto \frac{\hbar}{4\pi c\xi} = \frac{\hbar\sqrt{\varepsilon_0\mu_0}}{4\pi\xi}$. It seems that the electron, the positron, and perhaps all the elementary particles, might also behave as quantum vacuum solitons.

The calculations also show that the field $\tilde{\Xi}_{0_{k\lambda}}$ is a real entity, capable of interacting with the electrons in matter since it is expressed as a function of photon creation and annihilation operators $a_{k\lambda}^+$ and $a_{k\lambda}$ respectively. An interaction Hamiltonian between the vacuum and the electrons can be defined in this way, in order to describe the spontaneous emission effect. The atomic level energy shifts can also be interpreted as a result of the vacuum radiation pressure upon the atomic orbitals without involving singularities or neglecting infinite quantities in the calculations. This representation coherently links the electromagnetic theory to QED, attributes well defined physical properties to single photons and to the vacuum state and shows the complementarity of the wave-particle nature of light, as required by the experimental evidence. These concepts open perspectives for new experiments in order to determine the value of ξ and investigate the spatial expansion of the single photon state.

Finally, with an objective point of view, we can frankly confess that at the present state of knowledge, our comprehension of the photon and its relationship to the vacuum is still unsatisfactory and the purpose of this manuscript is to raise questions and give hints of answers, in order to stimulate further theoretical and experimental investigations.

Index

www.ingramcontent.com/pod-product-compliance
Lightning Source LLC
Chambersburg PA
CBHW050631190326
41458CB00008B/2224